Fundamentals of Public Utilities Management

Fundamentals of Public Utilities Management

Frank R. Spellman

CRC Press
Taylor & Francis Group
Boca Raton London New York

CRC Press is an imprint of the
Taylor & Francis Group, an **informa** business

First edition published 2021
by CRC Press
6000 Broken Sound Parkway NW, Suite 300, Boca Raton, FL 33487-2742

and by CRC Press
2 Park Square, Milton Park, Abingdon, Oxon, OX14 4RN

© 2021 Taylor & Francis Group, LLC

CRC Press is an imprint of Taylor & Francis Group, LLC

Reasonable efforts have been made to publish reliable data and information, but the author and publisher cannot assume responsibility for the validity of all materials or the consequences of their use. The authors and publishers have attempted to trace the copyright holders of all material reproduced in this publication and apologize to copyright holders if permission to publish in this form has not been obtained. If any copyright material has not been acknowledged please write and let us know so we may rectify in any future reprint.
Except as permitted under U.S. Copyright Law, no part of this book may be reprinted, reproduced, transmitted, or utilized in any form by any electronic, mechanical, or other means, now known or hereafter invented, including photocopying, microfilming, and recording, or in any information storage or retrieval system, without written permission from the publishers.

For permission to photocopy or use material electronically from this work, access www.copyright.com or contact the Copyright Clearance Center, Inc. (CCC), 222 Rosewood Drive, Danvers, MA 01923, 978-750-8400. For works that are not available on CCC please contact mpkbookspermissions@tandf.co.uk

Trademark notice: Product or corporate names may be trademarks or registered trademarks, and are used only for identification and explanation without intent to infringe.

Library of Congress Cataloging in Publication Data

Names: Spellman, Frank R., author.
Title: Fundamentals of public utilities management / Frank R. Spellman.
Description: First edition. | Boca Raton, FL : CRC Press, 2021. | Includes bibliographical references and index.
Identifiers: LCCN 2020020210 (print) | LCCN 2020020211 (ebook) | ISBN 9780367544393 (hardback) | ISBN 9781003089599 (ebook)
Subjects: LCSH: Water quality management. | Water-supply. | Sewage--Purification. | Water utilities--Management.
Classification: LCC TD365 .S6398 2021 (print) | LCC TD365 (ebook) | DDC 363.6/1068--dc23
LC record available at https://lccn.loc.gov/2020020210
LC ebook record available at https://lccn.loc.gov/2020020211

ISBN: 978-0-367-54439-3 (hbk)
ISBN: 978-1-003-08959-9 (ebk)

Typeset in Times
by Deanta Global Publishing Services, Chennai, India

Contents

Preface ix
Author Biography xv
Acronyms and Abbreviations xvii

Chapter 1 What Is Public Utility Management? 1

    Introduction 1
        Key Terms and Definitions 1
    Public Utilities Manager Qualifications 4
    A Regulatory View of Effective Management 8
        Attributes of Effectively Managed Water Sector Utilities 8
    Measuring Organization Effectiveness 10
        Benchmarking 13
    References and Recommended Reading 17

Chapter 2 Current Management Issues in Water and Wastewater
Treatment Operations 19

    Introduction 19
    The Paradigm Shift 20
        A Change in the Way Things Are Understood and Done ... 21
    Multiple-Barrier Concept 24
        Multiple-Barrier Approach: Wastewater Operations 25
    Management Problems Facing Water and Wastewater
    Operations 27
        Compliance with New, Changing, and Existing Regulations 27
        Maintaining Infrastructure 28
        Privatization and/or Re-engineering 32
    Water the New Oil? 33
        Water Use 35
        Technical Management versus Professional Management ... 37
    References and Recommendations 46

Chapter 3 Water/Wastewater Infrastructure: Energy Efficiency and
Sustainability 49

    Introduction 49
    Sustainable Water/Wastewater Infrastructure 51
        Cash Cows or Cash Dogs? 51
    The Water/Wastewater Infrastructure Gap 53
        Energy Efficiency: Water/Wastewater Treatment 53
    References and Recommended Reading 54

Chapter 4   Characteristics of Wastewater and Drinking Water Industries 55

Introduction (USEPA 2002) 55
Wastewater Treatment 55
Drinking Water Treatment 55
    The Honeymoon Is Over 56
Characteristics of Wastewater Industry 56
Wastewater Treatment Process: The Model 56
Characteristics of the Drinking Water Industry 57
Capital Stock and Impact on Operations/Maintenance 59
    Useful Life of Assets 59
    Operating and Maintaining (O&M) Capital Stock 60
    Wastewater Capital Stock 61
Drinking Water Capital Stock 63
    Costs of Providing Service 64
References and Recommended Reading 65

Chapter 5   Planning for a Sustainable Energy Future 67

Wastewater and Drinking Water Treatment: Energy Usage 67
    Present and Future Challenges 67
    Fast Facts 68
Benchmark It! 69
    Collection of Baseline Data and Tracking Energy Use 70
    Baseline Audit 74
    Field Investigation 75
    Create Equipment Inventory and Distribution of Demand
    and Energy 75
References and Recommended Reading 76

Chapter 6   Energy Efficient Operating Strategies 77

Introduction 77
Electrical Load Management 77
    Rate Schedules 78
    Energy Demand Management 78
    Electrical Load Management Success Stories 79
Operation and Maintenance: Energy Cost-Saving Procedures 82
    Chandler Municipal Utilities, Arizona (February 21, 2012) 83
    Airport Water Reclamation Facility, Prescott, Arizona
    (February 21, 2012) 84
    Somerton Municipal Water, Arizona (February 21, 2012) 85
    Hawaii County Department of Water Supply
    (February 21, 2012) 86
    Eastern Municipal Water District, California
    (February 21, 2012) 87

Contents vii

    Port Drive Water Treatment Plant, Lake Havasu, Arizona
    (February 21, 2012) 88
    Truckee Meadows Water Authority, Reno, Nevada
    (February 21, 2012) 90
    Tucson Water, Arizona (February 21, 2012) 92
    Chino Water Production Facility, Prescott, Arizona
    (February 21, 2012) 93
    Somerton Municipal Wastewater Treatment Plant, Arizona
    (February 21, 2012) 94
  References and Recommended Reading 95

Chapter 7 Energy Conservation Measures for Wastewater Treatment 97

  Introduction 97
  Pumping Energy Conservation Measures 98
    Pumping System Design 99
    Pump Motors 101
    Power Factor 106
    Variable Frequency Drives 106
  Aeration Blowers 108
    High-Speed Gearless (Turbo) Blowers 111
    Single-Stage Centrifugal Blowers 112
    New Diffuser Technology 112
    Innovative and Emerging Energy Conservation Measures 113
    UV Disinfection 114
    Membrane Bioreactors (MBRs) 119
    Anoxic and Anaerobic Zone Mixing 120
  References and Recommended Reading 123

Chapter 8 Digital Network Security 127

  Introduction 127
  The Digital Network 131
  The Bottom Line 133
  References and Recommended Reading 134

Chapter 9 SCADA 135

  What Is SCADA? 135
  SCADA Applications 136
  SCADA Vulnerabilities 138
    The Increasing Risk 141
    Adoption of Technology with Known Vulnerabilities 141
    Cyber Threats to Control Systems 143
    Securing Control Systems 144

	Steps to Improve SCADA Security	144
	21 Steps to Increase SCADA Security (DOE, 2001)	144
	References and Recommended Reading	150
Chapter 10	IT Security Action Plan	153
	IT Security Action Items	153
	Policy Development and Management	154
	Scams and Fraud	155
	Network Security	158
	Web Security	160
	Email	165
	Mobile Devices	167
	Employees	169
	Operational Security	172
	Payment Cards	174
	Digital Network Security	176
	The Bottom Line	178
	References and Recommended Reading	182
Chapter 11	Plant Security	185
	Introduction	185
	Consequences of 9/11	186
	Security Hardware Devices	190
	Physical Asset Monitoring and Control Devices	190
	Water Monitoring Devices	215
	Communication and Integration	224
	Cyber Protection Devices	227
	Adoption of Technologies with Known Vulnerabilities	230
	Cyber Threats to Control Systems	232
	Securing Control Systems	233
	References and Recommended Reading	233
Glossary		235
Index		249

Preface

Management development is neither new nor unique to the public sector. Why then has it become a hot issue? In general, communities are finding that there is a gap between how their public sector utilities are and how the interests of the community need them to be now or in the future. Community leaders are finding something missing between existing public service cultures and the public interest. A common complaint is lack of dedication to the underlying values of public service and the interests of the citizens served. A common response seems to be the attempt to promote a certain kind of manager.

The most important role of managers has been to solve the problems and challenges faced in a specific environment. In this text, the specific environment addressed is public sector utility management—with specific emphasis placed on modern water/wastewater utilities. The current economic environment in public sector utility management is one of uncertainty and change. The U.S. economy has shown signs of limited recovery since mid-1999 but seems to be expanding at the present time. Even with economic expansion, local government, operating under ever-increasing fiscal constraints and diminishing resources, is forced to reduce expenditures and limit expansion.

Good management is a critical component of good public service. When a public sector utility worker—as with almost any other worker in the general workforce—is promoted to any level of management, certain benefits accrue with the promotion. In many cases, these promotions include increased pay, increased benefits, increased pride and self-satisfaction, and increased responsibilities. It is this last "benefit," increased responsibility, that is often seen as more of an obligation than a benefit. For many newly promoted managers, increased responsibility is most difficult to accept and to effect.

Let's state this same opinion in a different manner: by asking a question.

When a public sector utility worker is promoted to any level of management, is the new manager, no matter how well motivated, properly equipped to assume the responsibility incumbent with the position? In many respects this question is relative, almost nebulous, because it depends on the selection process and the qualifications of the worker. This is a universal problem, of course, faced by all those who make new worker-to-management selections. The key part of this question is "how well equipped" is the new manager? The answer, of course, will vary. However, if the new manager is not properly prepared, the transition can be analogous to running in a minefield while blindfolded.

To be effective and successful at the rank-and-file (worker) level, public sector workers need only follow directions from the supervisor and perform work output at a level of effectiveness approved by the supervisor. The worker's focus is on performing to a narrow standard of accomplishment, which, in turn, can result in job security. Thus, the typical public sector worker is less likely to give much thought to current and pressing issues facing the overall industry that are not part and parcel of his/her daily work routine; that is, their view is narrow, not holistic. This is the case even though many of these current and pressing issues may directly or indirectly impact the worker's employment future in the public sector job.

There is another public sector utility management level that needs to be addressed—mid-level management. Mid-level manager jobs in utilities are usually filled by college graduates who are specialists (based on educational achievements) in narrow fields. For example, in water treatment it is not unusual, and actually quite normal, to hire biologists, environmental scientists, chemists, and engineers to fill various technical positions. In electrical and natural gas utility work, engineering graduates are commonly hired. Engineers and/or hydrologists usually fill mid-level management positions in stormwater management. Experience has shown that all of these technical specialists are needed in public sector utility work. The problem arises when technical specialists who are not experienced in management are expected to manage. This is not to say that a biologist or chemist or engineer cannot manage. Instead, it may be difficult (and sometimes disastrous) if an engineer or biologist or other technical expert is required to perform the dual function of technical expert and manager, especially when that person is not properly trained or suited to manage. Based on well-documented examples, this problem is more often the case than the exception. One thing is certain: graduation from college does not automatically qualify one to manage.

In order to manage effectively, the manager must not only have the ability and experience required to manage (to lead) but also be cognizant of the problems facing public sector utilities. These management problems are many and varied. Consider, for example, one area of concern. During the 1980s and the 1990s a wave of new buzzwords, *reengineering*, *privatizing*, and *rightsizing*, swept the developed and developing world, affecting the many public sector operations partially or fully privatized/reengineered/right-sized in the transition economies. The transition from the old way to the new way of doing public sector utility business is still evolving, but at a rapid rate.

The question is why? Why this apparent rush to change, especially in developed countries? Publicly owned and operated utilities (i.e., water, wastewater, electric, natural gas, stormwater management, and, to some degree, public transportation, telecommunications, and others) have been around for a long time (the exception being stormwater management and telecommunications, which are relatively recent). Many utilities have operated well and accomplished their intended missions with little or no public controversy. In most cases, when the public demands water at the tap, a waste-emptied toilet bowl after the flush, electricity at a throw of the on-switch, likewise for natural gas, and controlled stormwater runoff, these services have been readily available. At least this is one view residing on one edge of a double-edged sword.

The other edge? The public, always a demanding customer, wants more and better quality from public sector utilities, which many perceive to be wasteful, ineffective, and in many cases just plain unresponsive to its needs. It is within this atmosphere that we have witnessed the rise of accountability and competition for public sector programs. Programs that are unable to justify their value are being downsized or eliminated altogether.

Preface

Again, what is the problem? Why change something that apparently works—to a degree at least? We all know that change is controversial, difficult, costly, troublesome, and often cumbersome. So why not leave well enough alone?

Sorting through the current arguments for and against change in public utility management and operations evokes an emotional firestorm that would test Solomon. Almost inevitably, the eye of the beholder colors conclusions. But a few facts seem clear.

In the United States, for example, state and local governments now face a series of unprecedented challenges: budget deficits, bloated workforces, decaying infrastructures, suffocating regulatory compliance, shrinking tax bases, citizen opposition to new taxes and facilities, and taxpayer (ratepayer) imposed tax (rates) and spending limitations. These problems are often compounded by the perception of ratepayers and politicians who view rate payments or budget expenditures to public sector utilities as nothing more than throwing good money at bad money—money that never seems to make a dent in filling the black hole.

For the typical water and wastewater operator, stormwater manager, and/or electric and natural gas worker (i.e., public utility rank-and-file or public utility person) and some mid-level managers, budget deficits, bloated workforces, decaying infrastructures, shrinking tax bases, and the other associated problems are as foreign as the man in the moon. These problems are not in their domain; they're someone else's problem—typically the managers' problem.

Unfortunately, such problems, while not necessarily germane to public utility-workers, only remain floating in the rarefied atmosphere occupied by those holding upper management positions. Accordingly, public utility persons may be aware of obvious utility shortcomings (a plant unit process that belongs in the bone yard along with the other dinosaurs) but have no power or inclination to effect change.

The worker's situation or view does change, of course, whenever he or she joins the upper echelons of management through promotion. These "new managers" are, unfortunately, often unsuited to take over the responsibilities incumbent with their new leadership status. Their new position and subsequent outlook might be analogous to the sideline cheerleader who is suddenly suited out in full football garb, thrown onto the playing field and expected to perform in a positive way in a team effort to win the game.

The problems facing utilities currently require a new breed of public sector manager. This new breed, inspired by the successful streamlining of American business, is trying to meet these challenges—not by increasing taxes or government spending—by fundamentally transforming government through a process called privatization, reengineering, or rightsizing.

Privatizing, reengineering, and/or rightsizing means establishing clear priorities and asking questions that successful utilities regularly ask, such as: If we were not doing this already, would we start? Is this activity central to our mission? What is our mission? Do we have the need for a mission? Does a central all-controlling authority better make decisions or are they better made by consensus management or by empowering the employees to share input on decision making? If we were to design this organization from scratch, given what we now know about modern technology, what would it look like?

Any current seasoned or newly advanced manager can ask these questions. The problem arises in attempting to answer them. Knowledge, awareness, thinking outside the box, learned management skill, and common sense all play a crucial role in answering these questions. Coming up with the "correct" answer(s) is even more challenging. This dilemma points to the need to pre-train prospective managers before they are promoted. Unfortunately, this process is usually more of an afterthought than common practice. Experience has shown that promotions within the public sector often are based on seniority—the next available senior worker—and not on potential management skill.

Today's public sector utility manager faced with privatization, reengineering, or rightsizing problems is also faced with many additional challenges. One of these major challenges struck home with the 9/11 terrorist attack. Many managers have found out the brutal way that security is no longer one of those backburner concerns—just a cumbersome necessity. Security is not only on the front burner now and not only a life-style-sustaining necessity but also a huge challenge for any manager, no matter how seasoned.

Complicating matters are even more managerial challenges. For example, complying with current, new, and changing regulations is not exactly a cakewalk in a pristine high-alpine meadow on a sunny spring day.

Because the challenges facing today's utility managers are too long to list, in this text we aim our focus on five areas of pressing concern to most water public sector managers. It should be noted, however, that though intentionally directed toward water and wastewater utilities, the information provided in this text applies to all utilities.

- Complying with regulations, and coping with new and changing regulations
- Maintaining infrastructure and conservation of energy
- Privatization, reengineering, and rightsizing
- Benchmarking
- Upgrading security

This text does not provide all the solutions or answers to the problems currently facing managers and new managers in the public sector. However, the text does provide information for constructing a roadmap for successful accomplishment and/or compliance with the listed items above. This is accomplished by describing each challenge, and by pointing out the many pitfalls that should be avoided in taking on the challenges.

Written primarily for new managers, this text also is designed to provide insight and information to managers who have been managing for some time. In this presentation the focus is not on the primary and traditional management skills of coordinating, organizing, and directing the workforce—many excellent texts are available on developing and enhancing these skills. Instead, the focus is on the five areas listed earlier as they relate to waterworks, wastewater and stormwater infrastructure, and energy conservation, all of which are germane to the current or new manager's challenges in managing utility operations in a competitive, performance-based manner critical to the current culture. At present, many available management texts ignore, avoid, or pay cursory lip service coverage to these important areas.

Though directed at public utility sector managers and prospective managers in water and wastewater operations, this production will serve the needs of students, teachers, consulting engineers, and technical personnel in city, state, and federal public sector employment. Moreover, in order to maximize the usefulness of the material contained, the text is written in straightforward, user-friendly, plain English and conversational style, a characteristic of the author's style; that is, a failure to communicate is not allowed and not even given a first, second, or third thought.

To assure correlation to modern practice and design, illustrative problems and case studies are presented throughout the content in terms of commonly used managerial parameters.

Finally, this text and its content are accessible to those who have no experience with public sector utility management. If you work through the presentation systematically, an understanding of and skill in management techniques can be acquired—adding a critical component to your professional knowledge.

Frank R. Spellman
Norfolk, Virginia

Author Biography

Frank R. Spellman, PhD, is a retired, full-time adjunct assistant professor of environmental health at Old Dominion University, Norfolk, Virginia, and the author of more than 149 books covering topics ranging from concentrated animal feeding operations (CAFOs) to all areas of environmental science and occupational health. Many of his texts are readily available online at Amazon.com and Barnes and Noble.com, and several have been adopted for classroom use at major universities throughout the United States, Canada, Europe, and Russia; two have been translated into Spanish for South American markets. Dr. Spellman has been cited in more than 950 publications. He serves as a professional expert witness for three law groups and as an incident/accident investigator at wastewater treatment facilities and/or incidents involved with MRSA contraction, supposedly due to contact with wastewater and fatalities occurring at treatment plants for the U.S. Department of Justice and a northern Virginia and Nebraska law firm. In addition, he consults on homeland security vulnerability assessments for critical infrastructures including water/wastewater facilities nationwide and conducts pre-Occupational Safety and Health Administration (OSHA)/Environmental Protection Agency EPA audits throughout the country. Dr. Spellman receives frequent requests to co-author with well-recognized experts in several scientific fields; for example, he is a contributing author of the prestigious text *The Engineering Handbook*, 2nd ed. (CRC Press). Dr. Spellman lectures on wastewater treatment, water treatment, and homeland security and lectures on safety topics throughout the country and teaches water/wastewater operator short courses at Virginia Tech (Blacksburg, Virginia). In 2011–12, he traced and documented the ancient water distribution system at Machu Pichu, Peru, and surveyed several drinking water resources in Amazonia-Coco, Ecuador. Dr. Spellman also studied and surveyed two separate potable water supplies in the Galapagos Islands; he also studied and researched Darwin's finches while in the Galapagos. He is a Certified Safety Professional (CSP), a Certified Hazardous Materials Manager (CHMM) and Certified Environmental Trainer (Industrial Hygiene). He holds a BA in public administration, a BS in business management, an MBA, and an MS and PhD in environmental engineering.

Acronyms and Abbreviations

°C	Degrees Centigrade or Celsius
°F	Degrees Fahrenheit
µ	Micron
A-C	Alternating Current
Al^3	Aluminum Sulfate (or Alum)
Amp	Amperes
APWA	American Public Works Association
AS	Activated Sludge
ASCE	American Society of Civil Engineers
ATM	Atmosphere
AWRF	Airport Water Reclamation Facility
AWWA	American Water Works Association
BAT	Best Available Technology
BEP	Best Efficiency Point
bhp	Brake Horsepower
BNR	Biological Nutrient Removal
BOD	Biochemical Oxygen Demand
BPR	Biological Phosphorus Removal
CCCSD	Central Contra Costa Sanitary District
CEC	California Energy Commission
CEE	Consortium for Energy Efficiency
CFO	Cost Flow Opportunity
COD	Chemical Oxygen Demand
cu feet	Cubic Feet
CWA	Clean Water Act
DBP	Disinfection By-product
D-C	Direct Current
DO	Dissolved Oxygen
DoD	Department of Defense
DOE	Department of Energy
DS	Distribution System
DSM	Demand Side Management
ECM	Energy Conservation Measure
EMWD	Eastern Municipal Water District
EPA	Environmental Protection Agency
EPACT	Energy Policy Act
EPRI	Electric Power Research Institute
$FeCl_3$	Ferric Chloride
GAO	Government Accountability Office
GHG	Greenhouse Gas
GPD	Gallons per Day
GPM	Gallons per Minute

GUI	Graphic User Interface
GWDR	Ground Water Disinfection Rule
GWh	Gigawatt Hour
HDWK	Headworks
HMI	Human Machine Interface
hp	Horsepower
HTTP	Hypertext Transfer Protocol
ICS-CERT	Industrial Control Systems Cyber Emergency Response Team
IDSs	Intrusion Detection Systems
I&I	Inflow and Infiltration
IOA	International Ozone Association
IUVA	International Ultraviolet Association
kW	Kilowatt
kWh	Kilowatt-Hour
kWh/year	Kilowatt-Hours per Year
LPHO	Low Pressure High Output
M	Mega
MBR	Membrane Bioreactor
MCC	Motor Control Center
MG	Million Gallons
MGD	Million Gallons per Day
mg/L	Milligrams per Liter (Equivalent to Parts per Million)
MLE	Modified Ludzack-Ettinger Process
MLSS	Mixed Liquor Suspended Solids
MMBTU	Metric Million British Thermal Units
MPN	Most Probable Number
MTCO$_2$	Metric Tons of Carbon dioxide Equivalent
MTU	Master Terminal Unit
MW	Megawatt
mv	Millivolt
N	Nitrogen
NACWCA	National Association of Clean Water Agencies
NAWC	National Association of Water Companies
NEMA	National Electrical Manufacturers Association
NEPA	National Environmental Policy Act
NIPC	National Infrastructure Protection Center
NO$_3$	Nitrate
NPDES	National Pollutant Discharge Elimination System
O&M	Operation and Maintenance
ORP	Oxidation-Reduction Potential
OSHA	Occupational Safety and Health Organization
OTE	Oxygen Transfer Efficiency
PAO	Phosphate Accumulating Organisms
PDA	Personal Digital Assistant
PDWTP	Port Drive Water Treatment Plant
PG&E	Pacific Gas and Electric

Acronyms and Abbreviations

PGP	Pretty Good Privacy
PLC	Programmable Logic Controller
POTWs	Publicly Owned Treatment Works
psi	Pounds per Square Inch
psig	Pounds per Square Inch Gauge
RAS	Return Activated Sludge
rpm	Revolutions per Minute
RTU	Remote Terminal Unit
SBR	Sequencing Batch Reactor
SCADA	Supervisory Control and Data Acquisition
SDWA	Safe Drinking Water Act
SRT	Solids Retention Time
SSL	Secure Socket Layer
SWTR	Surface Water Treatment Rule
SQL	Sequence Query Language
TCCR	Total Coliform Rule
TDH	Total Dynamic Head
TEP	Tuscon Electric Power
THM	Trihalomethane
TMWA	Truckee Meadows Water Authority
TSS	Total Suspended Solids
TQM	Total Quality Management
UV	Ultraviolet Light
UVT	UV Transmittance
VA	Volt Ampere
VFD	Variable Frequency Drive
VOCs	Volatile Organic Chemicals
VPN	Virtual Private Network
VSS	Volatile Suspended Solids
W	Watt
WAS	Waste Activated Sludge
WEF	Water Environment Federation
WEFTEC	Water Environment Federation Technical Exhibition & Conference
WERF	Water Environment Research Foundation
WLAN	Wireless Local Area Network
WPA2	Wi-Fi Protected Access 2
WPCP	Water Pollution Control Plant
WSU	Washington State University
WTP	Water Treatment Plants
WWTP	Wastewater Treatment Plant
XML	Extensible Markup Language

1 What Is Public Utility Management?

The most important arrow in any manager's quiver can be summed up in one word:

Credibility, Credibility, Credibility.

A leader [or manager] is best when people barely know he exists, when his work is done, his aim fulfilled, they will say: we did it ourselves.

Lao Tzu

INTRODUCTION

Public utility management, also known as public works management, oversees the operations, maintenance, safeguarding, and preservation of the systems responsible for water treatment, wastewater collection and treatment, and distribution of safe potable water to a localized community. Keep in mind, however, that public utilities also can include energy, telecommunications, and other local services; in spite of this, in this book our focus is water service. With regard to actual hands on or eye on operations, the public utilities manager supervises the staff, interacts (networks) with the public (ratepayers and stakeholders), is the principal guardian of ratepayers and other contributors' funds, keeps an eye on plant and employee safety and health, stays in good standing with regulators—meets with liaisons from environmental and health agencies and with other government entities to discuss various issues affecting public utilities management—and coordinates with other municipal officials to ensure the efficient and professional function of facilities and the timely and reliable distribution of resources. The general and daily tasks a utility manager performs on a routine basis are shown in Figure 1.1.

With regard to the public utility manager's constant and never-ending focus beyond routine tasks, his or her center of attention is usually directed mainly in five areas. These main areas of primary focus, as shown in Figure 1.2, consist of staff, ratepayers, regulators, plant security, and the organization's governing body.

KEY TERMS AND DEFINITIONS

To study any aspect of wastewater and drinking water treatment management operations, you must master the language associated with the technology. Each technology has its own terms with its own accompanying definitions. Many of the terms used in water/wastewater treatment are unique; others combine words from many different technologies and professions. One thing is certain: water/wastewater managers

FIGURE 1.1 Routine tasks performed by utility managers.

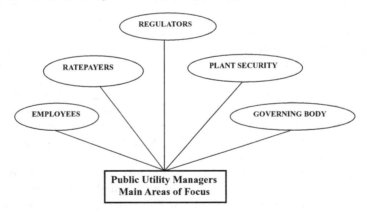

FIGURE 1.2 Public utility managers' main areas of focus.

without a clear understanding of the terms related to their profession are ill-equipped to perform their duties in the manner required. Although this text includes a glossary of terms at the end of the work, we list and define many of the terms used right up front, the ones relative to each chapter. Experience has shown that an early introduction to keywords is a benefit to readers. An up-front introduction to key terms facilitates a more orderly, logical, systematic learning activity. Those terms not defined in this section are defined as they appear in the text.

What Is Public Utility Management?

Key Terms Used in This Chapter

The following terms are presented in this chapter or are introduced here for a better understanding of the chapter's contents. These definitions provide a brief overview of their meaning (USEPA, 2008).

- **Attribute**: A characteristic or outcome of a utility that indicates effective performance.
- **Benchmarking**: Often used as a measurement tool in comparing similar operations and processes by measuring across organizations and/or sectors to identify best practices, set improvement targets, and simply to measure progress.
- **Effective utility management**: Management that improves products and services, increases community support, and ensures a strong and viable utility into the future.
- **Gap analysis**: Defining the present state of an enterprise's operations, the desired or "target" state, and the gap between them.
- **Internal trend analysis**: Comparison of outcomes or outputs relative to goals, objectives, baselines, targets, and standards.
- **Life-cycle cost**: The total of all internal and external costs associated with a product, process, or activity throughout its entire life cycle—from raw materials acquisition to manufacture/construction/installation, operation and maintenance, recycling, and final disposal.
- **Performance measurement**: Evaluation of current status and trends; it can also include a comparison of outcomes or outputs relative to goals, objectives, baselines, targets, standards, other organizations' performance or processes (typically called benchmarking), etc.
- **Operations and maintenance expenditure**: Expenses used for day-to-day operation and maintenance of a facility.
- **Operating revenue**: Revenue realized from the day-to-day operations of a utility.
- **Performance measure**: A particular value or characteristic designated to measure input, output, outcome, efficiency, or effectiveness.
- **Standard operating procedure**: A prescribed procedure to be followed routinely; a set of instructions having the force of a directive, covering those features of operations that lend themselves to a definite or standardized procedure without loss of effectiveness.
- **Strategic plan**: An organization's process of defining its goals and strategy for achieving those goals. Often entails identifying an organization's vision, goals, objectives, and targets over a multi-year period of time, as well as setting priorities and making decisions on allocating resources, including capital and people, to pursue the identified strategy.
- **Stewardship**: The careful and responsible management of something entrusted to a designated person or entity's care; the responsibility to properly utilize its resources, including its people, property, and financial and natural assets.

- **Sustainability**: The use of natural, community, and utility resources in a manner that satisfies current needs without compromising future needs or options.
- **Watershed health**: The ability of ecosystems to provide the functions needed by plants, wildlife, and humans, including the quality and quantity of land and aquatic resources.

PUBLIC UTILITIES MANAGER QUALIFICATIONS

What are the qualifications to be a public works manager? A better question is what are the qualifications needed to be an *effective* public works manager? Well, if you attend college to obtain a degree in management with the intention of finding a management job in the public utility or public works field you will be exposed to a wide assortment of management theories, concepts, models, systems, ideas, and philosophies. So, the question becomes which of all of these concepts is the right one for becoming enlightened and qualified on how to manage a public utility?

Good question. The truth is you will probably be exposed to a wide range of administrative theories including traditional standard bearer: Fayol's well-known Five Functions of Management. The Five Functions of Management are:

1. Planning
2. Organizing
3. Commanding
4. Coordinating
5. Controlling

Although Fayol's Five Functions of Management have been around since 1916 and many more current concepts and theories of management have been developed and employed since, Fayol's Five Functions of Management are still relevant to organizations today. In a nutshell these five functions focus on the relationship between personnel and management. Basically, they provide points of reference so that problems can be solved in an innovative and productive manner.

In addition to exposure to and learning about Fayol's concepts in college, in many management courses on principles, functions, theories and so forth, you may also be exposed to group dynamics, Max Weber's Bureaucratic Theory, McGregor's Theory X and Theory Y, Taylor's Scientific Management, Fayol's 14 Principles of Management, and/or French and Raven's Five Forms of Power.

The next question is which of these management theories is the best, is the most important, is something the student must learn and must abide by? Students should be exposed to as many different management theories and practices as possible. Based on experience, it is best to generalize one's college education with a little of this and a little of that along with the required courses needed to complete training and to graduate.

Again, based on experience, when a newly graduated college student with a degree in management or other area related to management completes his or her area of study and graduates and enters the workforce, that person is not a manager. Being a college graduate is not a sure-fire element needed to manage, to be a manager.

What Is Public Utility Management?

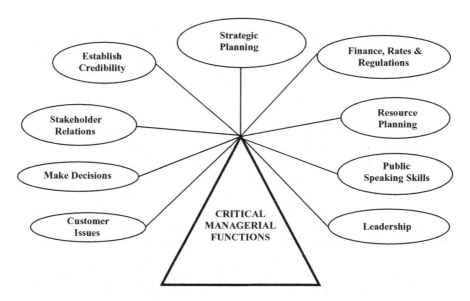

FIGURE 1.3 Critical managerial functions within the public utility organization.

What does it take to be a manager? What does it take to be an effective manager? A college degree in management opens the door for a new employee to become a manager but does not guarantee success. Success is accomplished via education plus on-the-job experience. Unless the new management graduate is a natural born leader, a General Patton, Stonewall Jackson, Napoleon, Genghis Khan, or someone of that flavor and natural ability, management must be learned; it must be groomed. Moreover, management must be generalized to the point whereby the prospective manager has the important mixture of managerial qualities including the physical, the mental, the moral, the technical, the education, and the experience to be an effective manager.

In the education and on-the-job training mode, what the new prospective manager is learning (hopefully) is how actual functions within the organization are integrated with one another. Learning these functions along with leadership ability and leadership style that does not include the Gestapo mentality is the actual ticket-punching that is required to become an effective manager. The important functions within the organization that are critical to success as a utility manager are shown in Figure 1.3.

Because they are important to managing and leading a public utility organization it is important to take a closer look at each of the critical managerial functions within the public utility organization.

- **Strategic Planning**—the public utility manager who is not far thinking, is not cognizant of emerging trends, and has no vision of utility strategy is like a person in the middle of the ocean without sails and oars and food and fresh water. In addition to being far thinking, the manager must write a strategic plan by exploring internal resources and capabilities against an external backdrop of regulation, economic factors, customer preferences, new technology, and competitive forces.

- **Establish Credibility**—when any manager directs his or her employees to perform this or that task and the manager has no idea what the workers' tasks are, there is a serious problem that develops with credibility. The quickest and surest way for any manager to lose respect and creditability is for that manager to not have a clue what he or she is talking about. So, what exactly is meant by a utility manager's lack of credibility? Well an example would be if a football coach has no idea what each player does on the field and what their responsibility is. This would be like a utility manager not understanding plant nomenclature and a given employee's responsibility—no knowledge of an employee's job description and requirements. Even more important regarding a utility manager's credibility is whether or not he or she understands the treatment plants' operation. Can the manager, for example, identify and explain what water disinfection is all about? If not, employees will have little to no respect for such a manager.
- **Stakeholder Relations**—the public utility manager must have a strong relationship with stakeholders. First, the public utility manager must find out who the stakeholders are. It is important to find out what the stakeholders' work styles, priorities, and specific project responsibilities are. The public utility manager must meet the stakeholders; the manager can't be a long-distance acquaintance. Instead, the manager must seek out the stakeholder and then keep close contact and listen, and listen, and, moreover, listen. Also, the public utility manager's interrelationship with the stakeholder(s) can be summed up in one word: Communicate, Communicate, Communicate.
- **Make Decisions**—have you ever worked for a manager, any manager at any level, who could not make a decision? Is there anything more frustrating? I think not. Many managers want to mull over decisions and there is nothing wrong with that, actually that is wise in most circumstances. However, the manager who can't or will not make a decision because he or she is afraid of failure is the failure itself. Being afraid to fall on one's sword is cowardice to the extreme. Managers must make decisions and live with the consequences. If the decision is a bad one, bone up to it and move on. No one respects a coward, period.
- **Customer Issues**—satisfying customer needs is what the public utility manager must focus on. Often times the public utility tends to forget who they are working for; it is their customers who they are working for. Customers have questions and needs. The public utility manager must address these questions and needs.
- **Finance, Rates, and Regulations**—any public utility manager who does not pay attention to finance, ratepayers' rates, and regulations is doomed and will lead his or her public utility to bankruptcy and into court for noncompliance regarding local, state and federal regulations. With regard to rates, any public utility manager must be sensitive to the rates charged to ratepayers; if the utility charges too much for its service(s) it will find itself at constant odds and conflict with the public it is designed to serve.
- **Resource Planning**—public utilities must not only plan ahead for whatever—chemicals, repair parts, equipment replacement, staffing but also

investments. A public utility manager must be trained in financial analysis used for investment analysis. The manager must understand investing equity and/or purchase of bonds to ensure the entity's financial success and stability. This financial acumen is critical to maintaining fair and equitable customer rates.

A Note About Staffing Wastewater Treatment Plants: A certain connotation, implication, suggestion, and basically a label or stereotype and definitely a false stigma exists about personnel (staff) that work in wastewater treatment plants and operations. Such a stigma is not rare. Think, for example, of exterminators, funeral and gravediggers, and toilet cleaners, and also parking inspectors, and animal control officers ... "Oh, how can you do that dirty work?" the unenlightened and open-minded average person might say.

Do you want a more expressive response from someone? Okay, simply say you are a sewage rat, or that you treat raw sewage or that you are a sewage operator. The immediate response:

Ugh!
Icky!
Oh, my god! Really?
Katie bar the door ... phew, let me out of here, right now.

These responses are commonly known or attributed to what is known as the "Yuck Factor." Now, it is true that most people feel that someone has to do it, but people can look down on that particular somebody who actually does the so-called dirty work. In contrast, if someone was to ask a water treatment plant operator what line of work he or she is in, the responses could be quite different and applied with a pat on the back and a comment stating, "Keep up the good work."

- **Public Speaking Skills**—if you have ever had to sit through a speaker's presentation that was presented in an evasive, vague, roundabout way in order to avoid responding to a question or making a definite statement, then you quickly understand the importance of being an effective speaker; one who can garner his or her audience's full attention and maintain it. The public utility manager who hems and haws around with his or her presentation will quickly lose his or her listeners. Instead of hemming and hawing, the speaker must adapt the presentation to the audience. The presentation must be of interest to the listeners. If the presentation is not spontaneous, it must be planned and practiced and full of energy and clarity. Presentations that include visual aids, humor, positives, quotes, or some remarkable facts are generally the most effective.
- **Leadership**—leadership is simply another word for managing. Public utility managers who can't lead by example, by doing, by presentation skills, and by force of personality will find leadership and/or management of people difficult.

A REGULATORY VIEW OF EFFECTIVE MANAGEMENT

After reviewing several explanations of what effective management of a public utility is, and based on personal experience of working with several different public utility managers in water and wastewater operations, I found that the best explanation detailing the attributes of effectively managed water service utilities is given by the United States Environmental Protection Agency (USEPA) in its 2008 *Effective Utility Management: A Primer for Water and Wastewater Utilities*. The USEPA developed its primer based on collaboration of the Water Environment Federation (WEF), National Association of Water Companies (NAWC), America Waterworks Association (AWWA), Association of Metropolitan Water Agencies, American Public Works Association (APWA), and National Association of Clean Water Agencies (NACWA). For anyone who aspires to become a public utility manager, this USEPA primer is highly recommended.

In USEPA's *Effective Utility Management: A Primer for Water and Wastewater Utilities* recommends ten attributes of effectively managed water sector utilities. These ten attributes are shown in Figure 1.4 and are explained below.

ATTRIBUTES OF EFFECTIVELY MANAGED WATER SECTOR UTILITIES

As stated earlier, USEPA (2008) has compiled 10 attributes of what they judge as effectively managed water sector utilities. In this section we briefly discuss each of the 10 attributes shown in Figure 1.4.

FIGURE 1.4 Ten attributes of effectively managed water sector utilities (USEPA, 2008).

Product Quality
Product quality refers to water operators tasked with producing water (the product) quality that is suitable and safe as potable water, produces a treated effluent, and produces process residuals in full compliance with regulatory and reliability requirements and consistent with customer, public health, and ecological needs.

Customer Satisfaction
Customer satisfaction deals with whether or not the water service provides reliable, responsive, and affordable services in line with explicit, customer-accepted service levels. Moreover, does the service provider provide timely feedback to maintain responsiveness to customer needs and emergencies?

Financial Viability
An effective public utility manager understands the full life-cycle cost of the utility and establishes and maintains an effective balance between long-term debt, asset values, operations and maintenance expenditures, and operating revenues. The effective manager establishes predictable rates—consistent with community expectations and acceptability—adequate to recover costs, provide for reserves, maintain support from bond rating agencies, and plan and invest for future needs.

Operational Optimization
An effective utility manager ensures ongoing, timely, cost-effective, reliable, and sustainable performance improvements in all facets of its operations. He or she minimizes resource use, loss, and impacts from day-to-day operations. The manager also maintains awareness of information and operational technology developments to anticipate and support timely adoption of improvements.

Infrastructure Stability
The effective utility manager understands the conditions of, and costs associated with, critical infrastructure assets; maintains and enhances the condition of all assets over the long-term at the lowest possible life-cycle cost; and takes on acceptable risks consistent with customer, community, and relation-supported service levels, as well as with anticipated growth and system reliability goals. In addition, an effective manager assures asset repair, rehabilitation, and replacement efforts are coordinated within the community to minimize disruptions and other negative consequences.

Operational Resiliency
An effective utility manager ensures utility leadership and staff work together to anticipate and avoid problems. The manager proactively identifies, assesses, and establishes tolerance levels for and effectively manages a full range of business risks (including legal, regulatory, financial, environmental, safety, security, and natural disaster-related) in a proactive way consistent with industry trends and system reliability goals.

Community Sustainability
Effective utility managers are explicitly cognizant of and attentive to the impacts their decisions have on current and long-term future community and watershed

health and welfare. He or she manages operations, infrastructure, and investments to protect, restore, and enhance the natural environment; efficiently uses water and energy resources; promotes economic vitality; engenders overall community improvement; and explicitly considers a variety of pollution prevention, watershed, and source water protection approaches as part of an overall strategy to maintain and enhance ecological and community sustainability.

Water Resource Adequacy

Effective public utility managers ensure water availability consistent with current and future customer needs through long-term resource supply and demand analysis, conservation, and public education.

Employee and Leadership Development

Effective public utility managers recruit a workforce that is competent, motivated, adaptive, and safety oriented. He or she establishes a participatory collaborative organization dedicated to continual learning and improvement and ensures employee institution knowledge is retained and improved upon over time. He or she also provides a focus on and emphasizes opportunities for professional and leadership development and strives to create an integrated and well-coordinated senior leadership team.

Stakeholder Understanding and Support

Effective utility managers engender understanding and support from oversight bodies, community and watershed interests, and regulatory bodies for service levels, rate structures, operating budgets, capital improvement programs, and risk management decisions. The manager actively involves stakeholders in the decisions that will affect them.

MEASURING ORGANIZATION EFFECTIVENESS

Measuring performance is one of the keys to utility management success. This section provides ideas about how to approach measurement and then offers measures for each attribute to help understand a utility's stats and progress.

You know there are some public utility managers who sit back in their padded chairs, pat themselves on the back, and think to themselves, "Gee, I am doing a wonderful job here."

Really?

Yes, I have personally met a few of them but when you ask them what is going on in their plant or plants, they seem to hem and haw and change the subject.

Have you ever witnessed such a response, such a manager?

Well, maybe the public utility manager's self-appreciation and pat on the back is warranted.

However, how does he or she know that? How did they compare their operation to others, to flagship operations, to any operation that is top notch, the best in class?

Or did they?

That is the question—if management does not conduct measurements of their operation, how are they to know how they are doing? How are they to know how they

What Is Public Utility Management?

compare to other like operations? If there are deficiencies, how are they to know that, unless they measure and correct the same?

Why should any public utility manager worry about plant performance? Well, job security might be one reason—that is, his or her job security. And then there is the trend toward privatizing public works facilities and operations in the hope of saving money and increasing efficiency.

Privatizing a public utility?

Yes, absolutely. It is a movement that is real and current and an aspect that management better be tuned into. Then there are those who might be too tuned in to the possibility of privatization. Consider the Waldrop Syndrome detailed in Case Study 1.1.

CASE STUDY 1.1 WALDROP SYNDROME

In 1995, G. (don't call me George) Daniel Waldrop, Director of Operations for one of the nation's largest sanitation districts, had a dilemma. The commission that governed his operation was bandying about the idea of privatizing the organization. The commissioners reasoned that public ratepayers would be better served by a private, efficient operation. In their view, private operation would not only save money but also bring in talented top managers [the commissioners felt that public service entities do not have a stellar history of attracting quality managers; for example, they felt their own operation was a repository for a bunch of good ol' boys, belonging to the ROAD (Retired On Active Duty) gang—unfortunately for the commissioners (and anyone else for that matter) once entrenched, it is almost impossible, excepting for criminal behavior bordering on murder, to get rid of such no-loads].

Waldrop understood that one of the first moves any privatizer would likely make would be to replace him, and many of his cronies, with proven private, non-public service professionals. This was, of course, unacceptable to him. However, Waldrop had one ace in the hole: He knew that the commission, before making the decision to privatize or not, had decided to conduct a study (a pilot study—engineers just love the words "pilot study") to determine if privatization actually made sense in their operation—in their case.

Waldrop understood that the results of the pilot study would help to determine whether privatization would actually occur. Knowing that the metrics involved with the commissioners' pilot study would focus on staffing, value-added operation, and total costs, Waldrop decided to nip the situation in the bud, so to speak. He knew each of the 14 wastewater treatment plants he was responsible for was "over-manned." Each plant had a workforce of approximately 44 personnel. Of this number, each plant had 4–6 plant operator assistants. Many of these plant operator assistants were licensed operators at the lowest professional certification levels (i.e., Class 4 or 3 Wastewater Operator: 4 being the lowest certification level—although some states like Virginia have created a Class 5 and Class 6 license level, both of which are lower classifications). Annual salaries for each assistant plant operator averaged around $32,000, depending on longevity. In reviewing assistant operator pay and experience levels, Waldrop was surprised to find that their average continuous time served on the job was 12 years. And while Waldrop understood that these operator assistants performed important functions at each of his treatment plants, they simply had to go. He reasoned that reducing plant staff levels to less than 40 would save a large outlay of annual funds for salaries, benefits, and personnel equipment (uniforms, safety shoes, training expenses, etc.). In regard to benefits (company paid life insurance, retirement, and total medical care), Waldrop multiplied the cost of these benefits, 1.4 times the annual salary, and determined that on average the plant operator assistants cost his operation $44,800 per annum. He then took the total number of assistants, 56, and multiplied 56 × $44,800 = $2,508,800. Waldrop was surprised at this total cost figure; he could not fathom how a bunch of assistant operators could cost so much!

Before the upcoming budget submission deadline (July 1, 2016), Waldrop subtracted the cost of operator assistants, $2,508,800, generated and distributed a memo giving all assistant treatment plant operators notice that their services were no longer required after June 30, 1996 (last day of the fiscal year), and sat back in his chair chain-smoking his cigarettes, actually quite pleased with himself.

Waldrop feeling good about his actions and himself was short-term, however. As it turned out, after the new fiscal year began July 1, 1996, the governor replaced the commission chairperson with a professional public administrator who had no intention (no personal motivation) for privatizing the district. So, the privatization effort evaporated before it ever got started.

The results of Waldrop's downsizing of treatment plant assistant operators did not sit well with the 14 plant managers who had lost their much-needed assistants. Waldrop, being the dysfunctional public service manager that he was, had not bothered to discuss his downsizing plans with anyone, let alone with the plant managers—the ones most affected by the downsizing. If Waldrop had taken the time and had the common decency (in the view of the plant managers) to discuss his downsizing plan with them, they would have adamantly and forcefully argued against such a short-sighted move. For example, the plant managers would have explained to Waldrop that the assistant operators formed a pool of fully trained personnel to instantly draw from whenever it became necessary to replace full-time treatment plant operators (the average turnover rate of full-time operators was approximately 3 per year per plant).

In addition to losing their pool of trained assistant operators to draw from to fill vacant positions, the lack of fully qualified assistant operators caused another glaring problem. The district had a far too generous annual leave policy that allowed employees with more than 20 years' service to take 6–8 weeks of annual leave, depending on pay grade and length of service. When the assistants were removed from the payroll, these natural fill-in personnel for those operators wanting to use their annual leave each year were no longer available.

Thus, it became increasingly harder for plant management personnel to allow key operators their annual leave each year.

Another unforeseen problem with doing away with the assistant operators was that when an operator became ill, injured, or encountered personal family problems, it was a huge burden on those plant personnel who had to fill the shoes of those who could no longer work. When the plants had a cadre of assistant plant operators to draw from to fill in for sick or otherwise not at work employees, this had not been a problem.

In summarizing the Waldrop Syndrome, it can be said that it adds meaning to Albert Einstein's thought in the following: "You cannot solve a problem with the mind-set that created it."

BENCHMARKING

One of the best ways for a public utility (or any other agency or business) to measure their effectiveness is to benchmark. As illustrated in Case Study 1.1, it is primarily out of self-preservation (to retain their lucrative positions) that many utility directors work against the trend to privatize water, wastewater, and other public operations. Usually the real work to prevent privatization is delegated to the individual managers in charge of each specific operation because they also have a stake in making sure that their relatively secure careers are not affected by privatization—it can be easily seen that working against privatization by these "local" managers is in their own self-interest and in the interest of their workers because their jobs may be at stake.

The question is, of course, how does one go about preventing his or her water and wastewater operation from being privatized? The answer is rather straightforward and clear: Efficiency must be improved at reduced cost of operations. In the real world, this is easier said than done—but is not impossible. For example, for those facilities under properly implemented and managed Total Quality Management (TQM), the process can be much easier.

The advantage TQM offers the plant manager is the variety of tools provided to help plan, develop, and implement water and wastewater efficiency measures. These tools include self-assessments, statistical process control, International Organization for Standards (ISO) 9000 and 14000, process analysis, quality circle, and benchmarking (see Figure 1.5).

In this text the focus is on the use of the benchmarking tool to improve water and wastewater operation's efficiency and is based on personal experience—characterized as a great experience. *Benchmarking* is a process of rigorously measuring your performance against best-in-class operations and using the analysis to meet and exceed the best in class.

Start---Plan---Research---Observe---Analysis---Adapt

FIGURE 1.5 Benchmarking process.

What is benchmarking?

1. Benchmarking against best practices gives water and wastewater operations a way to evaluate their operations overall.
   a. how effective
   b. how cost effective
2. Benchmarking shows plants both how well their operations stack up and how well those operations are implemented.
3. Benchmarking is an objective-setting process.
4. Benchmarking is a new way of doing business.
5. Benchmarking forces an external view to ensure correctness of objective-setting.
6. Benchmarking forces internal alignment to achieve plant goals.
7. Benchmarking promotes teamwork by directing attention to those practices necessary to remain competitive.

Potential results of benchmarking:

1. Benchmarking may indicate direction of required change rather than specific metrics.
   a. costs must be reduced
   b. customer satisfaction must be increased
   c. return on assets must be increased
   d. improved maintenance
   e. improved operational practices
2. Best practices translate into operational units of measure.

Targets:

- Consideration of available resources converts benchmark findings to targets.
- A target represents what can realistically be accomplished in a given time frame.
- Progress toward benchmark practices and metrics can be shown.
- Quantification of precise targets should be based on achieving benchmark.

Note: Benchmarking can be performance based, process based, or strategy based and can compare financial or operational performance measures, methods or practices, or strategic choices.

Benchmarking: The Process

When forming a benchmarking team, the goal should be to provide a benchmark that evaluates and compares privatized and re-engineered water and wastewater treatment operations to your operation in order to be more efficient, remain competitive, and make continual improvements. It is important to point out that benchmarking is more than simply setting a performance reference or comparison; it is a way to facilitate learning for continual improvements. The key to the learning process is looking outside one's own plant to other plants that have discovered better ways of achieving improved performance.

What Is Public Utility Management?

Benchmarking steps

As shown in Figure 1.5, the benchmarking process consists of five steps.

1. Planning—managers must select a process (or processes) to be benchmarked. A benchmarking team should be formed. The process of benchmarking must be thoroughly understood and documented. The performance measure for the process should be established (i.e., cost, time, and quality).
2. Research—information on the best-in-class performer must be determined through research. The information can be derived from the industry's network, industry experts, industry and trade associations, publications, public information, and other award-winning operations.
3. Observation—the observation step is a study of the benchmarking subject's performance level, processes, and practices that have achieved those levels, and other enabling factors.
4. Analysis—in this phase, comparisons in performance levels among facilities are made. The root causes for the performance gaps are studied. To make accurate and appropriate comparisons, the comparison data must be sorted, controlled for quality, and normalized.
5. Adaptation—this phase is putting what is learned throughout the benchmarking process into action. The findings of the benchmarking study must be communicated to gain acceptance, functional goals must be established, and a plan must be developed. Progress should be monitored and, as required, corrections in the process made.

Note: Benchmarking should be interactive. It should also recalibrate performance measures and improve the process itself.

CASE STUDY 1.2   BENCHMARKING: AN EXAMPLE

To gain better understanding of the benchmarking process, the following limited example (it is in outline and summary form only—discussion of a full-blown study is beyond the scope of this text) is provided.

RACHEL'S CREEK SANITATION DISTRICT
Introduction

In January 2007, Rachel's Creek Sanitation District formed a benchmarking team with the goal of providing a benchmark that evaluates and compares privatized and re-engineered wastewater treatment operations to Rachel's Creek operations in order to be more efficient and to remain competitive. After three months of evaluating wastewater facilities using the benchmarking tool, our benchmarking is complete. This report summarizes our findings and should serve as a benchmark by which to compare and evaluate Rachel's Creek Sanitation District operations.

Facilities

Forty-one wastewater treatment plants throughout the United States

The benchmarking team focused on the following target areas for comparison:

1. Re-engineering
2. Organization
3. Operations and maintenance
   a. Contractual services
   c. Materials and supplies
   d. Sampling and data collection
   e. Maintenance
4. Operational directives
5. Utilities
6. Chemicals
7. Technology
8. Permits
   a. Water quality
   b. Solids quality
   c. Air quality
   d. Odor quality
9. Safety
10. Training and development
11. Process
12. Communication
13. Public relations
14. Reuse
15. Support services
    a. Pretreatment
    b. Collection systems
    c. Procurement
    d. Finance and administration
    e. Laboratory
    f. Human resources

Summary of Findings:

Our overall evaluation of Rachel's Creek Sanitation District as compared to our benchmarking targets is a good one; that is, we are in good standing as compared to the forty-one target facilities we benchmarked with. In the area of safety, we compare quite favorably. Only plant 34, with its own full-time safety manager, appeared to be better than we are. We are very competitive with the privatized plants in our usage of chemicals and far ahead of many public plants. We are also competitive in the use of power. Our survey of what other plants are doing to cut power costs showed that we clearly identified those areas of improvement and our current effort to further reduce power costs is on track. We are far ahead in the optimization of our unit processes, and we are leaders in the area of odor control.

There were also areas that we needed to improve. To the Rachel's Creek employee, re-engineering applied to only the treatment department and had been limited to cutting staff, while plant practices and organizational practices were outdated and inefficient. Under the re-engineering section of this report, we have provided a summary of re-engineering efforts at the re-engineered plants visited. The experiences of these plants can be used to improve our own re-engineering effort. Next is our organization and staffing levels. A private company could reduce the entire treatment department staff by about 18–24 percent. The 18–24 percent are based on the number of employees and not costs. In the organization section of this report, organizational models and their staffing levels are provided as guidelines toward improving our organization and determining optimum staffing levels. The last big area that we need to improve is in the way we accomplish the work we perform. Our people are not used efficiently because of outdated and inefficient policies and work practices. Methods to improve the way we do work are found throughout this report. We noted that efficient work practices used by private companies allow plants to operate with small staffs.

Overall, Rachel's Creek Sanitation District's treatment plants are much better than other public service plants. Although some plants may have better equipment, better technology, and cleaner effluents, the costs in labor and materials are much higher than ours. Several of the public plants were in bad condition. Contrary to popular belief, the privately operated plants had good to excellent operations. These plants met permit, complied with safety regulations, maintained plant equipment, and kept the plant clean. Due to their efficiency and low staff, we felt that most of the privately operated plants were better than us. We agreed this needs to be changed. Using what we learned during our benchmarking effort, we can be just as efficient as a privately operated plant and still maintain our standards of quality.

The Bottom Line on Measurement

Measuring performance is one of the keys to public utility management success. Performance measurement can be conducted in-house by selected teams of employees who would be assigned to evaluate current internal utility performance status and trends. Based on personal experience I have found that benchmarking is the way to go when it comes to comparing your utilities' performance with others. This comparison should only be made with those utilities that are known for their effectiveness and best practices. Determining how the other "guy" does business compared to your agency can be a real eye opener and provide a roadmap for conducting operation and business at a higher level.

REFERENCES AND RECOMMENDED READING

Spellman, F.R. 2013. *Handbook of Water and Wastewater Treatment Plant Operations*. 3rd ed. Boca Raton, FL: Lewis Publishers.

USEPA. 2008. *Effective Utility Management: A Primer for Water and Wastewater Utilities*. Washington, DC: United States Environmental Protection Agency.

2 Current Management Issues in Water and Wastewater Treatment Operations

> The failure to provide safe drinking water and adequate sanitation services to all people is perhaps the greatest development failure of the twentieth century.
>
> **Gleick (1998, 2000)**

INTRODUCTION

Although not often thought of as a commodity (or, for that matter, thought about at all), water is a commodity—a very valuable, vital commodity. We consume water, waste it, dump waste into it, discard it, pollute it, poison it, and relentlessly modify the hydrological cycles (natural and urban cycles), with total disregard to the consequences: "too many people, too little water, water in the wrong places and in the wrong amounts. The human population is burgeoning, but water demand is increasing twice as fast" (De Villiers 2000). It is our position that with the passage of time, potable water will become even more valuable. Moreover, with the passage of even more time, potable water will be even more valuable than we might ever imagine—possibly (likely) comparable in pricing, gallon for gallon, to what we pay for gasoline, or even more—remember, we can live without gasoline but not without water. From urban growth to infectious disease and newly identified contaminants in water (e.g., forever chemicals), greater demands are being placed on our planet's water supply (and other natural resources). As the global population continues to grow, people will place greater and greater demands on our water supply (U.S. News 2000). The fact is—simply, profoundly, without a doubt in the author's mind—water is the new oil.

Earth was originally allotted a finite amount of water—we have no more or no less than that original allotment today—the earth is not making any more of it. Thus, it logically follows that, in order to sustain life as we know it, we must do everything we can to preserve and protect our water supply. Moreover, we also must purify and reuse the water we presently waste (i.e., wastewater).

> **DID YOU KNOW?**
>
> More than 50 percent of Americans drink bottled water occasionally or as their major source of drinking water—an astounding fact given the high quality and low cost of tap water in the United States.

THE PARADIGM SHIFT

Historically, the purpose of water supply systems has been to provide clean drinking water that is free of disease-causing organisms and toxic substances. In addition, the purpose of wastewater treatment has been to protect the health and well-being of our communities. Water/wastewater treatment operations have accomplished this goal by (1) prevention of practices that could lead to disease or nuisance, (2) avoidance of contamination of water supplies and navigable waters, (3) maintenance of clean water for survival of fish, bathing and recreation, and (4) generally, conservation of water quality for future use.

The purpose of water supply systems and wastewater treatment processes has not changed. However, there has been a paradigm shift primarily because of new regulations that include (1) protection against protozoan and virus contamination; (2) implementation of the multiple-barrier approach to microbial control; (3) new requirements of the Ground Water Disinfection Rule (GWDR), the Total Coliform Rule (TCR) and Distribution System (DS), the Lead and Copper (Pd/Cu) rule; (4) regulations for trihalomethanes (THMs) and Disinfection By-Products (DBPs); and (5) new requirements to remove even more nutrients (nitrogen and phosphorus) from wastewater effluent. We discuss this important shift momentarily but first it is important to abide by Voltaire's advice; that is, "If you wish to converse with me, please define your terms."

For those not familiar with the term "paradigm," it can be defined in the following ways. A **paradigm** is the consensus of the scientific community: "concrete problem solutions that the profession has come to accept" (Holyningen-Huene 1993). Thomas Kuhn coined the term "paradigm." He outlined it in terms of the scientific process. He felt that "one sense of paradigm, is global, embracing all the shared commitments of a scientific group; the other isolates a particularly important sort of commitment and is thus a subset of the first" (Holyningen-Huene 1993). The concept of paradigm has two general levels. The first is the encompassing whole, the summation of parts. It consists of the theories, laws, rules, models, concepts, and definitions that go into a generally accepted fundamental theory of science. Such a paradigm is "global" in character. The other level of paradigm is that it can also be just one of these laws, theories, models, etc., that combine to formulate a "global" paradigm. These have the property of being "local." For instance, Galileo's theory that the earth rotated around the sun became a paradigm in itself, namely a generally accepted law in astronomy. Yet, on the other hand, his theory combined with other "local" paradigms in areas such as religion and politics to transform culture. Paradigm can also be defined as a pattern or point of view that determines what is seen as reality.

We use the latter definition in this text.

A **paradigm shift** is defined as a major change in the way things are thought about, especially scientifically. Once a problem can no longer be solved in the existing paradigm, new laws and theories emerge and form a new paradigm, overthrowing the old if it is accepted. Paradigm shifts are the "occasional, discontinuous, revolutionary changes in tacitly shared points of view and preconceptions" (Daly 1980). Simply, a paradigm shift represents "a profound change in the thoughts, perceptions, and values that form a particular vision of reality" (Capra 1982). For our purposes, we use the term "paradigm shift" to mean a change in the way things are understood and done.

A Change in the Way Things Are Understood and Done

In water supply systems, the historical focus, or traditional approach, has been to control turbidity, iron and manganese, taste and odor, color, and coliforms. New regulations provided new focus, and thus a paradigm shift. Today, the traditional approach is no longer sufficient. Providing acceptable water has become more sophisticated and costly.

In order to meet the requirements of the new paradigm, a systems approach must be employed. In the systems approach, all components are inter-related. What affects one impacts others. The focus has shifted to multiple requirements (i.e., new regulations require the process to be modified or the plant upgraded).

To illustrate the paradigm shift in the operation of water supply systems, let us look back on the traditional approach of disinfection. Disinfection was used to destroy harmful organisms in water. While disinfection is still used for this purpose, it is now only one part of the **multiple-barrier approach**. Moreover, disinfection has traditionally been used to treat for coliforms only. Currently, because of the paradigm shift, disinfection now (and in the future) is used against coliforms, *Legionella*, *Giardia*, *Cryptosporidium*, and others. [**Note**: To effectively remove the protozoans Giardia and Cryptosporidium filtration is required; disinfection is not effective against the oocysts of Cryptosporidium.] Another example of the traditional vs. current practices is seen in the traditional approach to particulate removal in water to lessen turbidity and improve aesthetics. Current practice is still to decrease turbidity to improve aesthetics but now microbial removal plus disinfection is practical.

Another significant factor that contributed to the paradigm shift in water supply systems was the introduction of the Surface Water Treatment Rule (SWTR) in 1989. SWTR requires water treatment plants to achieve 99.9 percent (3 log) removal activation/inactivation of Giardia and 99.99 percent (4 log) removal/inactivation of viruses. SWTR applies to all surface waters and groundwaters under direct influence (GWUDI).

As mentioned earlier, the removal of excess nutrients such as nitrogen and phosphorus in wastewater effluent is now receiving more attention from regulators (USEPA) and others. One of the major concerns is over the appearance of dead zones in various water bodies (i.e., excess nutrients cause oxygen-consuming algae to grow and thus create oxygen-deficient dead zones). For example, in recent years it has not been uncommon to find several dead zone locations in the Chesapeake Bay region; consider the case study below.

CASE STUDY 2.1

The following newspaper article, written by the author, appeared in the January 5, 2005, *The Virginian-Pilot*. It is an Op Ed rebuttal to the article referenced in the text below. It should be pointed out that this article was well received by many but a few stated that it was nothing more than a rhetorical straw man. Of course, in contrast, I felt that the organizational critics were using the rhetorical Tin Man approach. That is, when you need to justify your cause and your organization's existence and need more grease, you squawk. The grease that many of these organizations require, however, is grease the consistency of paper cloth and is colored green—thus they squawk quite often. You be the judge.

CHESAPEAKE BAY CLEANUP: GOOD SCIENCE VS. "FEEL GOOD" SCIENCE

In your article, "Fee to help Bay faces anti-tax mood" (Va. Pilot, 1/2/05), you pointed out that environmentalists call it the "Virginia Clean Streams Law." Others call it a "flush tax." I call the environmentalist's (and others) view on this topic a rush to judgment, based on "feel good" science vs. good science. The environmentalists should know better.

Consider the following:

Environmental policymakers in the Commonwealth of Virginia came up with what is called the Lower James River Tributary Strategy on the subject of nitrogen (a nutrient) from the Lower James River and other tributaries contaminating the Lower Chesapeake Bay Region. When in excess, nitrogen is a pollutant. Some "theorists" jumped on nitrogen as being the cause of a decrease in the oyster population in the Lower Chesapeake Bay Region. Oysters are important to the local region. They are important for economic and other reasons. From an environmental point of view, oysters are important to the Lower Chesapeake Bay Region because they have worked to maintain relatively clean Bay water in the past. Oysters are filter-feeders. They suck in water and its accompanying nutrients and other substances. The oyster sorts out the ingredients in the water and uses those nutrients it needs to sustain its life. Impurities (pollutants) are aggregated into a sort of ball that is excreted by the oyster back into the James River.

You must understand that there was a time, not all that long ago (maybe 50 years ago), when oysters thrived in the Lower Chesapeake Bay. Because they were so abundant, these filter-feeders were able to take in turbid Bay water and turn it almost clear in a matter of three days. (How could anyone dredge up, clean, and then eat such a wonderful natural vacuum cleaner?)

Of course, this is not the case today. The oysters are almost all gone. Where did they go? Who knows?

The point is that they are no longer thriving, no longer colonizing the Lower Chesapeake Bay Region in numbers than they did in the past. Thus, they are no longer providing economic stability to watermen; moreover, they are no longer cleaning the Bay.

Ah! But don't panic! The culprit is at hand; it has been identified. The "environmentalists" know the answer—they say it has to be nutrient contamination; namely, nitrogen is the culprit. Right?

Not so fast.

A local sanitation district and a university in the Lower Chesapeake Bay region formed a study group to formally, professionally, and scientifically study this problem. Over a five-year period, using Biological Nutrient Removal (BNR) techniques at a local wastewater treatment facility, it was determined that the effluent leaving the treatment plant and entering the Lower James River consistently contained below 8 mg/L nitrogen (a relatively small amount) for five consecutive years.

The first question is: Has the water in the Chesapeake Bay become cleaner, clearer because of the reduced nitrogen levels leaving the treatment plant?

The second question is: Have the oysters returned?

Answer to both questions, respectively: no; not really.

Wait a minute. The environmentalists, the regulators, and other well-meaning interlopers stated that the problem was nitrogen. If nitrogen levels have been reduced in the Lower James River, shouldn't the oysters start thriving, colonizing, and cleaning the Lower Chesapeake Bay again?

You might think so, but they are not. It is true that the nitrogen level in the wastewater effluent was significantly lowered through treatment. It is also true that a major point source contributor of nitrogen was reduced with a corresponding decrease in the nitrogen level in the Lower Chesapeake Bay.

If the nitrogen level has decreased, then where are the oysters?

A more important question is: What is the real problem?

The truth is that no one at this point and time can give a definitive answer to this question.

Back to the original question: Why has the oyster population decreased?

One theory states that because the tributaries feeding the Lower Chesapeake Bay (including the James River) carry megatons of sediments into the bay (stormwater runoff, etc.), they are adding to the Bay's turbidity problem. When waters are highly turbid, oysters do the best they can to filter out the sediments but eventually they decrease in numbers and then fade into the abyss.

Is this the answer? That is, is the problem with the Lower Chesapeake Bay and its oyster population related to turbidity?

Only solid, legitimate, careful scientific analysis may provide the answer.

One thing is certain: before we leap into decisions that are ill-advised, that are based on anything but sound science, and that "feel" good, we need to step back and size up the situation. This sizing-up procedure can be correctly accomplished only through the use of scientific methods.

> Don't we already have too many dysfunctional managers making too many dysfunctional decisions that result in harebrained, dysfunctional analysis—and results?
>
> Obviously, there is no question that we need to stop the pollution of Chesapeake Bay.
>
> However, shouldn't we replace the timeworn and frustrating position that "we must start somewhere" with good common sense and legitimate science?
>
> The bottom line: We shouldn't do anything to our environment until science supports the investment. Shouldn't we do it right?
>
> **Frank R. Spellman**

MULTIPLE-BARRIER CONCEPT

On August 6, 1996, during the Safe Drinking Water Act Reauthorization signing ceremony, President Bill Clinton stated:

> A fundamental promise we must make to our people is that the food they eat and the water they drink are safe.

No rational person could doubt the importance of the promise made in this statement.

The Safe Drinking Water Act (SDWA), passed in 1974 and amended in 1986 and (as stated above) re-authorized in 1996, gives the United States Environmental Protection Agency (USEPA) the authority to set drinking water standards. This document is important for many reasons, especially because it describes how the USEPA establishes these standards.

Drinking water standards are regulations that the USEPA sets to control the level of contaminants in the nation's drinking water. These standards are part of the Safe Drinking Water Act's *multiple barrier approach* to drinking water protection (see Figure 2.1). As shown in Figure 2.1, the multiple-barrier approach includes the following elements:

1. **Assessing and protecting drinking water sources**—means doing everything possible to prevent microbes and other contaminants from entering water supplies. Minimizing human and animal activity around our watersheds is one part of this barrier.
2. **Optimizing treatment processes**—provides a second barrier. This usually means filtering and disinfecting the water. It also means making sure that the people who are responsible for our water are properly trained and certified and knowledgeable of the public health issues involved.
3. **Ensuring the integrity of distribution systems**—this consists of maintaining the quality of water as it moves through the system on its way to the customer's tap.
4. **Effecting correct cross-connection control procedures**—this is a critical fourth element in the barrier approach. It is critical because the greatest

Water and Wastewater Treatment 25

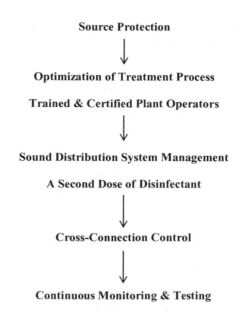

FIGURE 2.1 Multiple-barrier approach.

potential hazard in water distribution systems is associated with cross-connections to nonpotable waters. There are many connections between potable and nonpotable systems—every drain in a hospital constitutes such a connection—but cross-connections are those through which backflow can occur (Angele 1974).

5. **Continuous monitoring and testing of the water before it reaches the tap**—monitoring water quality is a critical element in the barrier approach. It should include having specific procedures to follow should potable water ever fail to meet quality standards.

With the involvement of the USEPA, local governments, drinking water utilities, and citizens, these multiple barriers ensure that the tap water in the United States and its territories is safe to drink. Simply, in the **multiple-barrier concept**, we employ a holistic approach to water management that begins at the source, continues with treatment, through disinfection and distribution.

MULTIPLE-BARRIER APPROACH: WASTEWATER OPERATIONS

Not shown in Figure 2.1 is the fate of the used water. What happens to the wastewater produced? Wastewater is treated via the multiple-barrier treatment train, which is the combination of unit processes used in the system. The primary mission of the wastewater treatment plant (and the operator/practitioner) is to treat the wastestream to a level of purity acceptable to return it to the environment or for immediate reuse (i.e., at the present time, reuse in such applications as irrigation of golf courses, etc.).

Water and wastewater professionals maintain a continuous urban water cycle on a daily basis. B.D. Jones (1980) sums this up as follows:

> Delivering services is the primary function of municipal government. It occupies the vast bulk of the time and effort of most city employees, is the source of most contacts that citizens have with local governments, occasionally becomes the subject of heated controversy and is often surrounded by myth and misinformation. Yet, service delivery remains the "hidden function" of local government.

DID YOU KNOW?

Artificially generated water cycles or the urban water cycles consist of (1) source (surface or groundwater), (2) water treatment and distribution, (3) use and reuse, and (4) wastewater treatment and disposition, as well as the connection of the cycle to the surrounding hydrological basins.

In *Fundamentals of Public Utility Management*, we focus on sanitary (or environmental) services (excluding solid-waste disposal)—water and wastewater treatment—because they have been and remain indispensable for the functioning and growth of cities. Water (next to air) is the most important life-sustaining product on Earth. Yet it is its service delivery (and all that it entails) that remains a "hidden function" of local government (Jones 1980). This "hidden function" is what this text is all about. We present our discussion in a completely new and unique dual manner—in what we call the new paradigm shift in water management and in the concept of the multiple-barrier approach. Essentially, in blunt, plain English the *Fundamentals of Public Utility Management* takes the "hidden" part out of services delivered by water and wastewater professionals.

Water service professionals provide water for typical urban, domestic, and commercial uses, eliminate waste, protect the public health and safety, and help control many forms of pollution. Wastewater service professionals treat the urban wastestream to remove pollutants before discharging the effluent into the environment. Water and wastewater treatment services are the urban circulatory system, the hidden circulatory system. In addition, like the human circulatory system, the urban circulatory system is less than effective if flow is not maintained. In a practical sense, we must keep both systems plaque-free and free-flowing.

Maintaining flow is what water and wastewater operations are all about. This seems easy enough: water has been flowing literally for eons—emerging from mud, rocks, silt, and the very soul of moving water, to carve a path, to pick up its load, to cargo its way to the open arms of a waiting sea where the marriage is consummated.

However, this is not to say that water and wastewater operations are without problems and/or challenges. After surviving the Y2K fiasco (were you surrounded by dysfunctional managers running around helter-skelter waiting until midnight?—as I was), the dawn of the 21st century brought with it, for many of us, aspirations of good things ahead in the constant struggle to provide quality food and water for humanity. However, the only way in which we can hope to accomplish this is to stay on the cutting edge of technology and to face all challenges head on. Some of these other challenges are addressed in the following sections.

MANAGEMENT PROBLEMS FACING WATER AND WASTEWATER OPERATIONS

Problems come and go, shifting from century to century, decade to decade, year to year, and from site to site. They range from the problems caused by natural forces (storms, earthquakes, fires, floods, and droughts) to those caused by social forces, currently including terrorism.

In general, five areas are of concern to many water and wastewater management personnel.

1. Complying with regulations, and coping with new and changing regulations
2. Maintaining infrastructure
3. Privatization and/or re-engineering
4. Benchmarking
5. Upgrading security

COMPLIANCE WITH NEW, CHANGING, AND EXISTING REGULATIONS

Note: This section is from the notes of Frank R. Spellman provided to J.E. Drinan (2001).

Adapting the workforce to the challenges of meeting changing regulations and standards for both water and wastewater treatment is a major concern. As mentioned, drinking water standards are regulations that the USEPA sets to control the level of contaminants in the nation's drinking water. Again, these standards are part of the SDWA's multiple-barrier approach to drinking water protection.

There are two categories of drinking water standards:

1. A **National Primary Drinking Water Regulation** (primary standard)—this is a legally enforceable standard that applies to public water systems. Primary standards protect drinking water quality by limiting the levels of specific contaminants that can adversely affect public health and are known or anticipated to occur in water. They take the form of Maximum Contaminant Levels or Treatment Techniques.
2. A **National Secondary Drinking Water Regulation** (secondary standard)—this is a non-enforceable guideline regarding contaminants that may cause cosmetic effects (such as skin or tooth discoloration) or aesthetic effects (such as taste, odor, or color) in drinking water. The USEPA recommends secondary standards to water systems but does not require systems to comply. However, states may choose to adopt them as enforceable standards. This information focuses on national primary standards.

Drinking water standards apply to public water systems which provide water for human consumption through at least 15 service connections, or regularly serve at least 25 individuals. Public water systems include municipal water companies, homeowner associations, schools, businesses, campgrounds, and shopping malls.

More recent requirements, for example, the Clean Water Act Amendments that came into effect in February of 2001, require water treatment plants to meet tougher standards, presenting new problems for treatment facilities to deal with, and offering some possible solutions to the problems of meeting the new standards. These regulations provide for communities to upgrade existing treatment systems and replace aging and outdated infrastructure with new process systems. Their purpose is to ensure that facilities are able to filter out higher levels of impurities from drinking water, thus reducing the health risk from bacteria, protozoa, and viruses, and that they are able to decrease levels of turbidity and reduce concentrations of chlorine by-products in drinking water.

With regard to wastewater collection and treatment, the National Pollution Discharge Elimination System (NPDES) program, established by the Clean Water Act, issues permits that control wastewater treatment plant discharges. Meeting permit is always a concern for wastewater treatment managers because the effluent discharged into water bodies affects those downstream of the release point. Individual point source dischargers must use the best available technology (BAT) to control the levels of pollution in the effluent they discharge into streams. As systems age, and best available technology changes, meeting permit with existing equipment and unit processes becomes increasingly difficult.

MAINTAINING INFRASTRUCTURE

During the 1950s and 1960s, the U.S. government encouraged the prevention of pollution by providing funds for the construction of municipal wastewater treatment plants, water pollution research, and technical training and assistance. New processes were developed to treat sewage, analyze wastewater, and evaluate the effects of pollution on the environment. In spite of these efforts, however, expanding population and industrial and economic growth caused pollution and resulting health problems to increase.

In order to make a coordinated effort to protect the environment, the National Environmental Policy Act (NEPA) was signed into law on January 1, 1970. In December of that year, a new independent body, the USEPA, was created to bring under one roof all of the pollution control programs related to air, water, and solid wastes. In 1972, the Water Pollution Control Act Amendments expanded the role of the federal government in water pollution control and significantly increased federal funding for the construction of wastewater treatment plants.

Many of the wastewater treatment plants in operation today are the result of federal grants made over the years. For example, because of the 1977 Clean Water Act Amendment to the Federal Water Pollution Control Act of 1972 and the 1987 Clean Water Act reauthorization bill, funding for wastewater treatment plants was provided.

Many large sanitation districts, with their multiple plant operations, and even a larger number of single plant operations in smaller communities in operation today

Water and Wastewater Treatment

are a result of these early environmental laws. Because of these laws, the federal government provided grants of several hundred million dollars to finance the construction of wastewater treatment facilities throughout the country.

Many of these locally or federally funded treatment plants are aging; based on my experience, I rate some as dinosaurs. The point is, many facilities are facing problems caused by aging equipment, facilities, and infrastructure. Complicating the problems associated with natural aging is the increasing pressure on inadequate older systems to meet the demands of increased population and urban growth. Facilities built in the 1960s and 1970s are now 50 to 60 years old, and not only are they showing signs of wear and tear, but they simply were not designed to handle the level of growth that has occurred in many municipalities.

Regulations often necessitate a need to upgrade. By matching funds or providing federal money to cover some of the costs, municipalities can take advantage of a window of opportunity to improve their facility at a lower direct cost to the community. Those federal dollars, of course, do come with strings attached; they are to be spent on specific projects in specific areas. On the other hand, many times, new regulatory requirements are put in place without the financial assistance needed to implement them. When this occurs, either the local community ignores the new requirements (until caught and forced to comply) or they face the situation and implement local tax hikes to pay the cost of compliance.

An example of how a change in regulations can force the issue is demonstrated by the demands made by Occupational Safety and Health Administration (OSHA) and the USEPA in their Process Safety Management (PSM)/Risk Management Planning (RMP) regulations (29 CFR 1910.119—OSHA). These regulations put the use of elemental chlorine (and other listed hazardous materials) under scrutiny. Moreover, because of these regulations, plant managers throughout the country are forced to choose which side of a double-edged sword cuts their way the most. One edge calls for full compliance with the regulations (analogous to stuffing the regulation through the eye of a needle). The other edge calls for substitution. That is, replacing elemental chlorine with a non-listed chemical (e.g., hypochlorite) or a physical (ultraviolet irradiation, UV) disinfectant—either way, a very costly undertaking.

Note: Many of us who have worked in water and wastewater treatment for years characterize PSM and RMP as the elemental chlorine killer. You have probably heard the old saying: "If you can't do away with something in one way, then regulate it to death."

Note: Changes resulting because of regulatory pressure sometimes mean replacing or changing existing equipment, increased chemical costs (e.g., substituting chlorine with hypochlorite typically increases costs threefold), and could easily involve increased energy and personnel costs. Equipment condition, new technology, and financial concerns are all considerations when upgrades or new processes are chosen. In addition, the safety of the process must be considered, of course, because of the demands made by the USEPA and OSHA. The potential of harm to workers, the community, and the environment are all under study, as are the possible long-term effects of chlorination on the human population.

CASE STUDY 2.2 CHESAPEAKE BAY AND NUTRIENTS: A MODEST PROPOSAL

Nutrient pollution in Chesapeake Bay and other water bodies is real and ongoing—the controversy over what is the proper mitigation procedure(s) is intense and never-ending (and very political). Nutrients are substances that all living organisms need for growth and reproduction. Two major nutrients, nitrogen and phosphorus, occur naturally in water, soil, and air. Nutrients are present in animal and human waste and chemical fertilizers. All organic material, such as leaves, and grass clippings contain nutrients. These nutrients cause algal growth and depletion of oxygen in the Bay, which leads to the formation of dead zones lacking in oxygen and aquatic life.

U.S. Fish and Wildlife Service (2007) points out that nutrients can find their way to the Bay from anywhere within the 64,000 square-mile Chesapeake Bay watershed—and that is the problem. All streams, rivers, and storm drains in this huge area eventually lead to the Chesapeake. The activities of over 13.6 million people in the watershed have overwhelmed the Bay with excess nutrients. Nutrients come from a wide range of sources, which include sewage treatment plants (20–22 percent), industry, agricultural fields, lawns, and even the atmosphere. Nutrient inputs are divided into two general categories: point sources and nonpoint sources.

Sewage treatment plants, industries, and factories are the major point sources. These facilities discharge wastewater containing nutrients directly into a waterway. Although each facility is regulated for the amount of nutrients that can be legally discharged, at times, violations still occur.

Wastewater treatment plants (point source dischargers) are of concern to us. It should be pointed out that wastewater treatment plants, approximately 350 outfall units that release effluents to nine major rivers and other locations all flowing into the Chesapeake Bay region, discharge around 20–22 percent of total nutrients into the Bay. Many target these point source dischargers as principal causes of oxygen depletion and creators of dead zone regions within the Bay. If this is true, one needs to ask the question: Why is wastewater discharged into the Chesapeake Bay in the first place? Water is the new oil, and if we accept this as fact we should therefore preserve and use our treated wastewater with great care and even greater utility. Thus, it makes good sense (to me) to reuse the wastewater that is presently discharged into Chesapeake Bay. This recycling of water saves raw water supplies in reservoirs and aquifers and limits the amount of wastewater that is discharged from wastewater treatment plants into public waterways, such as Chesapeake Bay. Properly treated wastewater could be used for many other purposes that raw water now serves, including irrigating lawns, parks, gardens, golf courses and farms; fighting fires; washing cars; controlling dust; cooling industrial machinery and towers and nuclear reactors; making concrete; and cleaning streets. It is my contention, of course, that when we get thirsty enough, we will find another use for properly treated and filtered wastewater, and when this occurs, we certainly will not use this treated wastewater for any other purpose than quenching our thirst. In reality, we are doing this already ... but this is the subject of another forthcoming book: *Water Is the New Oil: Sustaining Freshwater Supplies*.

Let's get back to the Chesapeake Bay problem.

The largest source of nutrients dumped into the Bay is from nonpoint sources. These nonpoint sources pose a greater threat to the Chesapeake Bay ecosystem, as they are much harder to control and regulate. It is my view that because of the difficulty of controlling runoff from agricultural fields and the lack of political will, wastewater treatment plants and other end-of-pipe dischargers have become targets of convenience for the regulators. The problem is that the regulators are requiring the expenditure of hundreds of millions to billions of dollars to upgrade wastewater treatment to biological nutrient removal, tertiary treatment and/or the combining of microfiltration membranes with a biological process to produce superior quality effluent—these requirements are commendable, interesting, achievable, but not at all necessary.

What is the alternative, the answer to the dead zone problem, the lack of oxygen problem in various locations in Chesapeake Bay? Simply, take a portion of the hundreds of millions of dollars earmarked for upgrading wastewater treatment plants (which is a total waste, in my opinion) and build mobile, floating platforms containing (electro-)mechanical aerators or mixers. These platforms should be outfitted with diesel generators and accessories to provide power to the mixers. The mixer propellers should be adjustable so they will be able to mix at a water depth of as little as 10 feet or as much as 35 feet. Again, these platforms are mobile. When a dead zone appears in the bay, the mobile platforms with their mixers are moved to a center portion of the dead zone area and energized at the appropriate depth—these mobile platforms are anchored to the Bay's bottom and so arranged to accommodate shipping and ensure that maritime traffic is not disrupted. The idea is to churn the dead zone water and sediment near the benthic zone and force a geyser-like effect above the surface to aerate the Bay water in the dead zone regions. Nothing adds more oxygen to water than natural or artificial aeration. Of course, while aerating and forcing oxygen back into the water, bottom sediments containing contaminants will also be stirred up and sent to the surface and temporary air pollution problems will occur around the mobile platforms. Some will view this turning up of contaminated sediments as a bad thing—not a good thing. I, on the contrary, point out that removing contaminants from the Bay by volatizing them is a very good thing.

How many of these mobile mixer platforms will be required? It depends on the number of dead zones. Enough platforms should be constructed to handle the average number of dead zones that appear in the Bay in the warm season.

Will this modest proposal—using aerators to eliminate dead zones—actually work? I do not have a clue. However, this proposal makes more sense to me than spending billions of dollars on upgrading wastewater treatment plants and effluent quality when this only accounts for 20–22 percent of the actual problem ... and the regulators and others do not have the political will or the brainpower to go after runoff, which is the real culprit in contaminating Chesapeake Bay with nutrient pollution.

Recall what the great mythical hero Hercules, the world's first environmental engineer, said: "dilution is the solution to pollution"—I agree with this. However, in this text, the solution to preventing dead zones in Chesapeake Bay is to prohibit discharge of wastewater from point sources (i.e., reuse to prevent abuse) and to aerate the dead zones.

PRIVATIZATION AND/OR RE-ENGINEERING

As mentioned, water and wastewater treatment operations are undergoing a new paradigm shift. I explained that this paradigm shift focused on the holistic approach to treating water. The shift is, however, more inclusive. It also includes thinking outside the box. In order to remain efficient and therefore competitive in the real world of operations, water and wastewater facilities have either bought into the new paradigm shift or been forcibly "shifted" to doing other things (often these "other" things have little to do with water/wastewater operations) (Johnson & Moore 2002).

Experience has shown that few words conger up more fear among plant managers than "privatization" or "re-engineering." *Privatization* means allowing private enterprise to compete with government in providing public services, such as water and wastewater operations. Privatization is often proposed as one solution to the numerous woes facing water and wastewater utilities, including corruption, inefficiencies (dysfunctional management), and the lack of capital for needed service improvements and infrastructure upgrades and maintenance. Existing management, on the other hand, can accomplish re-engineering, internally, or it can be (and usually is) used during the privatization process. *Re-engineering* is the systematic transformation of an existing system into a new form to realize quality improvements in operation, system capability, functionality, performance, or evolvability at a lower cost, schedule, or risk to the customer.

Many on-site managers consider privatization and/or re-engineering schemes threatening. In the worst-case scenario, a private contractor could bid the entire staff out of their jobs. In the best case, privatization and/or re-engineering are often a very real threat that forces on-site managers into workforce cuts (the Waldrop Syndrome, explained in case study 1.1), improving efficiency and cutting costs. While at the same time, on-site managers work to ensure the community receives safe drinking water and the facility meets standards and permits, with fewer workers—and without injury to workers, the facility, or the environment.

There are a number of reasons causing local officials to take a hard look at privatization and/or re-engineering.

1. **Decaying Infrastructures**—many water and wastewater operations include water and wastewater infrastructures that date back to the early 1900s. The most recent systems were built with federal funds during the 1970s, and even these now need upgrading or replacing. The USEPA

recently estimated that the nation's 75,000+ drinking water systems alone would require more than $100 billion in investments over the next 20 years. Wastewater systems will require a similar level of investment.
2. **Mandates**—the federal government has reduced its contributions to local water and wastewater systems over the past 30 years, while at the same time imposing stricter water quality and effluent standards under the Clean Water Act and Safe Drinking Water Act. Moreover, as previously mentioned, new unfunded mandated safety regulations, such as OSHA's Process Safety Management and USEPA's Risk Management Planning, are expensive to implement using local sources of revenues or state revolving loan funds.
3. **Hidden Function**—earlier we stated that much of the work of water/wastewater treatment is a "hidden function." Because of this lack of visibility, it is often difficult for local officials to commit to making the necessary investments in community water and wastewater systems. Simply, the local politicians lack the political will—water pipes and interceptors are not visible and not perceived as immediately critical for adequate funding. Thus, it is easier for elected officials to ignore them in favor of expenditures of more visible services, such as police and fire. Additionally, raising water and sewage rates to cover operations and maintenance is not always effected because it is an unpopular move for elected officials to make. This means that water and sewer rates do not adequately cover the actual cost of providing services in many municipalities.

In many locations throughout the United States, expenditures on water and wastewater services are the largest facing local governments today. (This is certainly the case for those municipalities struggling to implement the latest stormwater and nutrient reduction requirements.) Thus, this area presents a great opportunity for cost savings. Through privatization, water/wastewater companies can take advantage of advanced technology, more flexible management practices, and streamlined procurement and construction practices to lower costs and make critical improvements more quickly.

In regard to privatization, the view taken in this text is that ownership of water resources, treatment plants, and wastewater operations should be maintained by the public (local government entities) to prevent a tragedy of the commons–like event [i.e., free access and unrestricted demand for water (or other natural resources) ultimately structurally dooms the resource through overexploitation by private interests]. However, because management is also a "hidden function" of many public service operations (e.g., water and wastewater operations), privatization may be a better alternative to prevent creating a home for dysfunctional managers and for ROAD gangers—Retired On Active Duty.

The bottom line: water and wastewater are commodities whose quantity and quality are much too important to be left in the hands of public authorities.

WATER THE NEW OIL?

Even though more than 70 percent of Earth is covered with water, only 3 percent is fit for human consumption, of which two-thirds is frozen and largely uninhabited ice caps and glaciers, leaving 1 percent available for consumption. The remaining 97 percent is saltwater, which cannot be used for agriculture or drinking. If all the earth's water fit in a quart jug, available fresh water would not equal a teaspoon. Thus, some (including me) would consider water the new oil.

Let's take a moment and make a few important points about water, the new oil.

Unless you are thirsty, in real need of refreshment, when you look upon a glass of water you might ask: Well, what could be more boring? The curious might wonder what the physical and chemical properties of water are that make it (the water in the glass) so unique and necessary for living things. Again, when you look at a glass of water, taste and smell it—well, what could be more boring? Pure water is virtually colorless and has no taste or smell. But the hidden qualities of water make it a most interesting subject.

When the uninitiated becomes initiated to the wonders of water, one of the first surprises is that the total quantity of water on Earth is much the same now as it was more than three or four billion years ago, when the 320+ million cubic miles of it were first formed—there is no more fresh water on Earth today than there was millions of years ago. The water reservoir has gone round and round, building up, breaking down, cooling, and then warming. Water is very durable, but remains difficult to explain, because it has never been isolated in a completely undefiled state.

Have you ever wondered what the nutritive value of water is? Well, the fact is water has no nutritive value. It has none; yet it is the major ingredient of all living things. Consider yourself, for example. Think of what you need to survive—just to survive. Food? Air? Iphone? Twitter? Water? Naturally, the focus of this text is on water and water service. Water is of major importance to all living things; in some organisms, up to 90 percent of their body weight comes from water. Up to 60 percent of the human body, 70 percent of the brain, and 90 percent of the lungs are composed of water. About 83 percent of our blood is water, which helps digest our food, transport waste, and control body temperature. Each day humans must replace 2.4 liters of water, some through drinking and the rest taken by the body from the foods eaten.

There wouldn't be any you, me, or Lucy the dog without the existence of an ample liquid water supply on Earth. The unique qualities and properties of water are what make it so important and basic to life. The cells in our bodies are full of water. The excellent ability of water to dissolve so many substances allows our cells to use valuable nutrients, minerals, and chemicals in biological processes.

Water's "stickiness" (from surface tension) plays a part in our body's ability to transport these materials all through ourselves. Carbohydrates and proteins that our bodies use as food are metabolized and transported by water in the bloodstream; no less important is the ability of water to transport waste material out of our bodies.

Water is used to fight forest fires. Yet we use water spray on coal in a furnace to make it burn better.

Chemically, water is hydrogen oxide. It turns out, however, on more advanced analysis to be a mixture of more than thirty possible compounds. In addition, all of its physical constants are abnormal (strange).

At a temperature of 2,900°C some substances that contain water cannot be forced to part with it. And yet others that do not contain water will liberate it when even slightly heated.

When liquid, water is virtually incompressible; as it freezes, it expands by an eleventh of its volume.

For the above stated reasons, and for many others, we can truly say that water is special, strange, and different.

Important Point: As mentioned, water is called the "universal solvent" because it dissolves more substances than any other liquid. This means that wherever water goes, either through the ground or through our bodies, it takes along valuable chemicals, minerals, and nutrients.

WATER USE

In the United States, rainfall averages approximately $4,250 \times 10^9$ gal (gallons) a day. About two-thirds of this returns to the atmosphere through evaporation directly from the surface of rivers, streams, and lakes, and transpiration from plant foliage. This leaves approximately $1,250 \times 10^9$ gal a day to flow across or through the earth to the sea.

USGS (2004) estimated that about 408 billion gallons of water per day (abbreviated Bgal/d) were withdrawn for all uses in the United States during 2000. This total has changed by less than 3 percent since 1985 as withdrawals have stabilized for the two largest uses—thermoelectric power and irrigation. Fresh groundwater withdrawals (83.3 Bgal/d (billion gallons per day)) during 2000 were 14 percent more than during 1985. Fresh surface-water withdrawals for 2000 were 262 Bgal/d, varying less than 2 percent since 1985.

About 195 Bgal/d, or 8 percent of all freshwater and saline-water withdrawals for 2000, were used for thermoelectric power. Most of this water was derived from surface water and used for once-through cooling at power plants. About 52 percent of fresh surface-water withdrawals and about 96 percent of saline-water withdrawals were for thermoelectric-power use. Withdrawals for thermoelectric power have been relatively stable since 1985.

Irrigation remained the largest use of freshwater in the United States and totaled 137 Bgal/d for 2000. Since 1950, irrigation has accounted for about 65 percent of total water withdrawals, excluding those for thermoelectric power. Historically, more surface water than groundwater has been used for irrigation. However, the percentage of total irrigation withdrawals from groundwater has continued to increase, from 23 percent in 1950 to 42 percent in 2000. Total irrigation withdrawals were 2 percent more for 2000 than for 1995, because of a 16 percent increase in groundwater withdrawals and a small decrease in surface-water withdrawals. Irrigated acreage more than doubled between 1950 and 1980, then remained constant before increasing by nearly 7 percent between 1995 and 2000. The number of acres irrigated with sprinkler and microirrigation systems has continued to increase and now comprises more than one-half the total irrigated acreage.

Public-supply withdrawals were more than 43 Bgal/d for 2000. Public-supply withdrawals during 1950 were 14 Bgal/d. During 2000, about 85 percent of the population in the United States obtained drinking water from public suppliers, compared to 62 percent during 1950. Surface water provided 63 percent of the total during 2000, in contrast to 74 percent in 1950.

Self-supplied industrial withdrawals totaled nearly 20 Bgal/d in 2000, or 12 percent less than in 1995 and 24 percent less than in 1985. Estimates of industrial water use in the United States were largest during the years from 1965 to 1980, but during 2000, estimates were at the lowest level since reporting began in 1950. Combined withdrawals for self-supplied domestic, livestock, aquaculture, and mining were less than 13 Bgal/d for 2000 and represented about 3 percent of total withdrawals.

California, Texas, and Florida accounted for one-fourth of all water withdrawals for 2000. The states with the largest surface-water withdrawals were California and Texas, which had large withdrawals for irrigation and thermoelectric power.

All this factual information is interesting. Well, it is interesting to those of us who are admirers, purveyors, and/or students of water. Obviously, these are the folks that read and use a book like this one. However, the question is, what does all this information have to do with water being the new oil?

Water is the new oil because there is no more fresh water on Earth today than there was millions of years ago. Yet at the present time, more than 6 billion people share it. Since the 1950s, the world population has doubled, and water use has tripled. A simple extrapolation of today's water usage compared to projected usage in the future shows that water will become a much more important commodity than it is today. Earlier it was stated that the day is coming when a gallon of water will be comparable in value to (or even more expensive than) a gallon of gasoline. There are those who will read this and will shake their heads in doubt and state: "Water is everywhere; water belongs to no one; water belongs to everyone; no one owns the water; water pours freely from the sky; water has no real value ... certainly water is nowhere near as valuable as gasoline ... nowhere near as valuable as gold or diamonds."

Water has no real value? Really?

In regard to water and diamonds and which of the two is more valuable, consider the following. Adam Smith, the 18th century philosopher credited with laying the foundations for modern economics in his epic book, *The Wealth of Nations*, described the paradox of diamonds and water. Smith asked: how could it be that water, so vital to life, is so cheap, while diamonds, used only for adornment, are very costly? Smith pointed out that when it comes to value, a container full of diamonds is exponentially more valuable than an equal amount of water. In today's value system (as it was in Smith's), this is true, of course. It is true unless you are dying of thirst. While on the edge of dying, what value would you place on that same container of diamonds? On that particular container of water? If you were offered one or the other, which would you choose? Which would you give up everything you own for? And that is my point. While Adam Smith used the paradox for his own pedagogical purposes (explaining the basic concepts of supply and demand and showing that prices reflect relative scarcity), today the paradox provides a troubling description of the way water is treated in our economy. While water may be critical to life itself, we don't have a clue as to its true value (EPA 2003). No, not yet ... we have not reached that point yet. However, with the majority of the world's population being relatively thirsty and many dying of thirst or dying from drinking filthy, pathogen-contaminated water, the dawn of new understanding is just around the corner. Moreover, as population continues to grow and degradation of the world's supply continues and global climate change accelerates, it is my view that diamonds and water will switch places. Diamonds will lose some value when compared to safe, potable drinking water. This will occur because when it comes to sustaining life and quenching our thirst, all the diamond-encrusted drinking glasses filled to the brim with crystal clear water will be just what the doctor ordered, thank you very much!

Years ago, when I first stated that water will be more valuable than an equal amount of gasoline, many folks (reviewers) asked me what part of the planet Mars I was from? Well, I have not been to Mars and have not changed my opinion on the ever-increasing value of water—and this same realization will soon confront us all. By the way, you know that water we flush down our toilets and drains? The day is coming, in my opinion, when we will have direct pipe-to-pipe connections from wastewater treatment plants to our municipal potable water supply systems. Why? Water is the new Oil. Furthermore, have you heard about the recent discovery of the ancient presence of water on Mars? My guess is that if we do not protect our water supplies, the Mars of today may be the Earth of tomorrow. This is a thought to keep close at hand, close at heart, very close to the brain cells ... as a reminder of what really matters.

If you do not accept the premise that water is the new oil, maybe you are willing to indirectly accept the point on this matter. That is, we can use water to make oil. No, although possible, I am not talking about hydrogen from water converted into fuel. Instead, consider that we can turn algae into fuel. Consider that the scientists at Old Dominion University (ODU) in Norfolk, Virginia, for example, are conducting successful research on growing algae in treated sewage and extracting fatty oils from the weedy slime, then converting the oils into cleaner-burning fuel. As part of the research project, the algae is grown in tanks at a wastewater treatment plant in Norfolk then converted to biofuel at an ODU facility. It should be pointed out that this wastewater-grown-algae-to-oil-to-fuel-process has already proven itself in New Zealand (Harper 2007).

The bottom line: The day is drawing near when water is the new oil. This day is closer than what we may be willing to readily acknowledge.

DID YOU KNOW?

Growing algae in wastewater will soak up nutrients in wastewater at the wastewater plant, thus helping the receiving water body, which could suffer from excessive nutrients discharged by such treatment plants.

TECHNICAL MANAGEMENT VERSUS PROFESSIONAL MANAGEMENT

Water treatment operations management is management that is directed toward providing water of the right quality, in the right quantity, at the right place, at the right time, and at the right price to meet various demands. Wastewater treatment management is directed toward providing treatment of incoming raw influent (no matter what the quantity), at the right time, to meet regulatory requirements, and at the right price to meet various requirements.

The techniques of management are manifold both in water resource management and wastewater treatment operations. In water treatment operations, for example, management techniques may include: storage to detain surplus water available at one time of the year for use later, transportation facilities to move water from one place to another, manipulation of the pricing structure for water to reduce demand, adoption of changes in legal systems to make better use of the supplies available, use of techniques to make more water available through watershed management, cloud seeding desalination of saline or brackish water, or area-wide educational programs to teach conservation or reuse of water (Mather 1984).

Many of the management techniques employed in water treatment operations are also employed in wastewater treatment. In addition, wastewater treatment operations employ management techniques that may include upgrading present systems for nutrient removal, reuse of process residuals in an earth-friendly manner, and area-wide educational programs to teach proper domestic and industrial waste disposal practices.

Whether managing a waterworks or wastewater treatment plant, the manager, in regard to expertise, must be a well-rounded, highly skilled individual. No one questions the need for incorporation of these highly trained practitioners—well-versed in the disciplines and practice of sanitary engineering, biology, chemistry, hydrology, environmental science, safety principles, accounting, auditing, technical aspects, and operations—in both professions. Based on years of experience in the water and wastewater profession and personal experience dealing with high-level public service managers, however, engineers, biologists, chemists, and others with no formal management training and no proven leadership expertise are often hindered (limited) in their ability to solve the complex management problems currently facing both industries. I admit my biased view in this regard because my experience in public service has, unfortunately, exposed me to more dysfunctional than functional managers.

So, what is dysfunctional management? How is it defined? Consider case study 1.5; maybe it will provide an answer.

CASE STUDY 2.3   DYSFUNCTIONAL MANAGEMENT

Earlier, in case study 1.1, G. Daniel Waldrop, Chief Operations Officer for a large, well-known sanitation district, took various steps to downgrade his workforce in an effort to economize operational expenses. The need to economize (in his mind anyway) was driven by a threat to privatize his operation. Such a move would have been a direct threat to Waldrop because he assumed that a private operation would have immediately replaced him and his cronies with competent, proven management, and less-expensive personnel. Simply, privatization of his organization was viewed by Waldrop, career-wise, as the ultimate kiss of death—"the" career killer.

It is interesting to note that when Waldrop was downsizing his operation by doing away with all the operator assistant positions, he was also looking at other, higher-up positions to downgrade to further reduce costs; in his view, he simply had to make deeper cuts in personnel (labor), the operation's largest expense. However, when it came to delivering bad news, especially to those engineers and other managers who might be the target of such cost-cutting measures, Waldrop was not the type to address the problem face to face. Waldrop suffered from that chronic jellyfish syndrome—like the stately jellyfish, Waldrop simply did not possess a backbone.

Waldrop hired an outside professional management firm with extensive experience in auditing organizations and determining where to make cuts in the organization to economize operations. The cost-cutter (hatchet-man) assigned to assess Waldrop's operation was highly skilled in his work; he had more than twenty years' experience in the field.

It is interesting to note that the cost-cutter needed no more than 10 minutes at each of Waldrop's 14 facilities to come up with the solution to Waldrop's problem.

Two weeks after being hired, the cost-cutter met with Waldrop to deliver his findings and recommendations. The cost-cutter began his presentation by telling Waldrop that he could immediately save his organization approximately $1,355,200.00 per year simply by doing away with plant manager positions. Approximately 12 years earlier, Waldrop had created plant manager positions to manage each plant. The main qualification for each of these plant managers was that they had to be college graduates and engineers. Prior to hiring the engineers, the plants had been managed for years by plant superintendents. The superintendents were blue-collar employees who had worked their way to the top and had managed the plants successfully for several years; they had proven themselves to be competent managers. Waldrop found that he had difficulty talking to and relating to these superintendents—the blue-collar types. Simply, he had nothing in common with any of them and they did not communicate at his level—his super-educated, superior intellect level. Thus, the plant manager position was created so Waldrop would have someone to deal with that could understand him. Further, Waldrop, a dysfunctional manager (but a smart one) knew that the only thing protecting him from blame and termination for anything that went wrong at the plants was the managers below him at whom who he could readily point the finger of blame. Waldrop was one of those dysfunctional managers who were expert in personal survival techniques. He kept a worn, faded, tattered, dog-eared copy of *The Prince* within close reach—at all times. Niccolò Machiavelli would have been proud.

The hired cost-cutter explained (and Waldrop, with jaw dropped, listened) that each plant manager was paid a salary of $88,000.00 per annum times 1.4 for benefits equaling $123,200.00 per annum times 11 (total number of managers) for a grand total of 1,355,200.00—a considerable cost that could be made into a considerable savings.

The cost-cutter ignored the shocked look on Waldrop's face and continued. "You presently have both a plant manager and a plant superintendent at each facility … this is overkill … your upper management is top heavy with plant managers, plant superintendents, chief operators, and lead operators—too much costly management. When I asked what function the plant managers performed at each site, I pretty much got the same answer from each." The cost-cutter stopped to check his notes and Waldrop, feeling severe chest pains and an irritable bowel problem at the same time, sat there like a potted plant, wilting with each of the cost-cutter's words. "The plant managers all basically say the same thing," the cost-cutter said. "Basically, they state that their job is to overlook operations but never to interfere with the plant superintendents ... unless the superintendents are incoherent or out of control for one reason or another. I asked them if that was their total function ... they all said no that they were assigned to various organizational teams studying various organizational problems in line with the organization's TQM program."

After swallowing hard and finally finding his voice, Waldrop said, "Ok, thanks for your information and please forward your expense invoice to me so that I can pay you ... we will no longer need your help ... ah, thank you, thank you very much."

And with those words of ultimate dismissal the cost-cutter departed, never to return. Waldrop just sat there in his chair shaking his head in disgust. "It will be a cold day in hell before I ever get rid of my brother and sister engineers ... no matter how much money it would save," he told himself.

In my Industrial Environmental Management, Risk Management, and Occupational Safety and Health Management undergraduate and graduate classes at Old Dominion University, my students are required to study and complete research projects on many aspects of professional management. Much of their research is based on their experiences while interning at public service entities where they spend their summers learning on-the-job skills and earning income to help pay their way through college.

Upon completing their internships and returning to the classroom, I am always amazed at how pumped up with information the students are. They can't wait to relate their internship experience to their classmates (actually, all intern students are required to formally present their experiences to their peers in my classes).

In the last eight years one of the interesting trends I have noticed about these returning students is that almost all of them express their thoughts about the managers they were exposed to in a negative light.

I find this trend interesting. I usually give the students a chance to say what they have to say (in my view, there is no such thing as an incorrect opinion) and then invariably, eventually I ask them: "Can you say anything positive about the folks you worked for?"

They usually seem surprised that I ask such a question but eventually answer that they did ... or that they had a few good experiences here and there. Moreover, it is interesting that most of them give the same answer to the following question: "What did you find was the biggest problem with the managers you worked with and for?"

Almost every respondent answered: "The managers I was exposed to seem to have difficulty in making decisions ... they simply did not want to make a decision."

Water and Wastewater Treatment

After listening to the students' presentations, I finally get around to my main question: "Do you think the organization you worked at during the summer is functional or dysfunctional?" Most reply that the organization is functional but that the managers seemed dysfunctional.

Year after year, I ask students, "How do you define the dysfunctional manager?" The answers I receive are usually vague, ambiguous, and definitely incomplete, like most OSHA regulations. One student told me she would rather define beauty (which is impossible, according to her), than define the dysfunctional manager.

Eventually, the students, as all students eventually do, turned to their teacher (me) for a definition of the dysfunctional manager. When this first occurred, for one of the few times I was tongue-tied ... I stumbled over explaining that the dysfunctional could not lead ... that he could not foresee the future ... that he was not a problem solver ... that he simply lacked that which is not so common: common sense. A few of the really good students accused me of copping out; they were correct, of course. I simply could not answer this question without having some time to think about it.

Keeping in mind that identifying a dysfunctional manager is relatively easy (especially if you have the misfortune of working for one) but that describing one is much harder, I thought about it and case study 2.4 is the result.

CASE STUDY 2.4   DEFINING THE DYSFUNCTIONAL MANAGER

Have you ever felt the need (the absolute, overwhelming, crushing, pulsating need) to get close and personal with your manager? That is, have you ever wanted to get in your manager's face and laugh, scream, howl, shout, spit, throttle, or just stare with that stare that sums *it* all up?

Maybe you are one of those straightforward folks who would take a different approach. Maybe you would just walk up to your manager and coolly, calmly, and, with intense purpose, reach out and place your resignation on his/her junk heaped desk. And then march off smartly to his/her office door, make an about face, and issue your former oppressor the ultimate *coup de grace* whose meaning cannot be misunderstood.

Some daring folks (maybe you are one) might take a more dramatic approach. Those with ice water flowing through their veins might terminate a disturbing and distressing relationship with the boss by walking right into his/her office, picking up his/her trash can, moving it over to the nearest wall, inserting one foot into the can, donning a gas mask, and then placing an index finger against the wall. At this point in time, the manager is either ignoring you or breathless with anticipation of your next move. This is what you want, of course—that is, the manager's full, undivided attention. When you have an attentive audience, you simply yell out as loud as you can: "ELEVATOR GOING UP—YOU WANT TO RIDE ALONG, NERD-BREATH?"

Of course, people have used many other ways to end their relationships with their bosses, their employers, their managers, their harassers—some classic, some more imaginative. The question becomes: What was it that drove you to end it, to quit, to self-terminate? Was your terminal visit to face the manager brought about by feelings of anxiety, panic, frustration, depression, grief, guilt, shame, worry, anger, jealousy, and belligerence? Or maybe what it really boiled down to was you were feeling like you were "going all to pieces." Certainly, it was some "event" or series of "events" that torque your jaws—this is a given. Having reached this state, you may have decided that it was time to put the pieces back together again. Thus, your actions were brought about for one simple reason (that same old standard reason): You had all you were going to take—and you were not going to take any more.

Well, whatever it was that initiated your "fed up with it all attitudes" (especially with your manager) and your decisive action, I understand. The fact is most of the rest of us out here in the real-world workplace (sometimes referred to as la-la land) also understand.

Now, it is probably true that there is a small minority of workers out there (a very small group—too small to count) that doesn't understand what it is that would drive any person to act in the manner or manners described above.

These brain-washed or brain-dead, disillusioned individuals probably see their managers (and management in general) as their guiding light, their knight in shining white armor, their Horatio at the bridge, the personification of Mother Teresa, their father or mother figure, or something even more stellar. This type of warped thinking is, of course, quite sad. You might even characterize it as dysfunctional thinking. But that is another story—another topic. What I am addressing in this particular treatise (diatribe) is dysfunctional management in the public service sector (actually, in any sector of employment). Although it should be quite obvious to all readers that dysfunctional thinking does have something to do with inept or dysfunctional management.

Don't you agree?

Let's get back to the reason(s) why the worker (aren't we all classified in this terrible, ignoble category) suddenly decides that he or she has had enough. That is, why the worker goes off the edge or what I call throwing off the jellyfish label. (**Note**: No disrespect or disparagement is intended or directed toward the amorphous, silk-like, stately jellyfish. Instead, the intent here is to point out in an analogous fashion that jellyfish are not equipped with backbones; instead, they are gelatinous masses that tend toward quivering and quavering with each rustle of the wind, with each lapping of the water surface, with each undulation of the watery mass—simply stated, jellyfish go with the flow; they have little choice—they are not equipped with backbones.)

In the process of throwing off the jellyfish label, some individual situation, some single event occurs—an initiator that causes or literally drives a suddenly enlightened worker to metamorphose from jellyfish to 600-pound gorilla mode. Why the worker finally finds the backbone and energy to sort things out, to right millions of wrongs, to change to something better is worth our consideration.

Don't you agree?

Exactly what generates the hurricane-force wind at the back of the worker who makes this metamorphosis is what this treatise is all about. More specifically, this treatise is an account of the types of dysfunctional management practices that have driven environmental compliance workers in public service to react, to respond, to leave their places of employment.

Actually, when you get right down to it, is it not the practice of dysfunctional management that motivates us to react at all times—in one way or another? Pain stimulus is a highly effective motivational technique, whether it's called behavior modification or torture (B.F. Skinner, go back to sleep; you have nothing positive to add to this discussion).

At this point in the presentation a few individuals out there (less than 3 percent of the readers, probably) are asking themselves: What exactly is a dysfunctional manager? In order to ensure that all readers have a clear understanding (an absolutely clear understanding) of what a dysfunctional manager is, a definition or explanation is called for. Remember Voltaire's ingenious statement?—"If you wish to converse with me, please define your terms." Simply stated, if management is considered to be the glue that holds an organization together, then functional management is the Super Glue. On the other hand, dysfunctional management is that agent (however nebulous it might be) that works to break the cohesive bond.

You don't care for that analogy? You still want a more precise definition? OK.

A more accurate, correct, and precise explanation of what a dysfunctional manager really is can be made more clearly by explaining what a dysfunctional manager *is not*. With this in mind, consider what I call the *dysfunctional dozen*.

THE DYSFUNCTIONAL DOZEN

1. A dysfunctional manager *is not* qualified to prevent turnover costs and hassles by using specific hiring and interviewing techniques.
2. A dysfunctional manager *is not* qualified to blend differing personality types, backgrounds, and age groups into a smooth-running, productive team.
   Are you starting to get the idea?
   Yeah, I thought so.
   Let's move on.
3. A dysfunctional manager *is not* qualified to supervise former peers and friends without losing their respect.
4. A dysfunctional manager *is not* qualified to establish boundaries for supervisor/subordinate relationships that will not be misunderstood.
5. A dysfunctional manager *is not* qualified to quickly identify difficult employees and redirect them with swiftness and ease.
6. A dysfunctional manager *is not* qualified to relay constructive criticism without it being taken personally—even by the least sensitive employee.
7. A dysfunctional manager *is not* qualified to originate project plans and set goals that his/her staff will buy into.

8. A dysfunctional manager *is not* qualified to control absenteeism and tardiness (hell, they have enough trouble controlling their own)—they don't try ... why should they?
9. A dysfunctional manager *is not* qualified to fire or take corrective action—or to learn the legal implications for each.
10. A dysfunctional manager *is not* qualified to work under pressure.
11. A dysfunctional manager *is not* qualified to organize people, projects, and schedules ... on an ongoing basis (actually, *not* on any basis).
12. A dysfunctional manager *is not* qualified to keep top performers at their maximum level without burning out.

After having reviewed the dysfunctional dozen, even those who are normally confused should now have an understanding of what I am talking about in this treatise: *dysfunctional management*, and the *dysfunctional public service manager*. In case there is still some lingering doubt, I will sum it up for you quite succinctly: the dysfunctional manager is without conscience, without heart, without feeling, without direction, without discretion, without motivation, without scruples, and without leadership ability.

Before moving on from the tenets of the "dozen," I'd like to point out that while some managers may fit cleanly into any one division of the dysfunctional dozen, most will display additional dysfunctional behaviors at one time or another ... sooner or later. It's a given. But you should keep in mind that it only takes "competence" in one member of the "dozen" to fully qualify as a dysfunctional manager.

Have you ever worked for a manager who meets the criteria listed in *each* member of the "dozen"?

If you have, you are not alone.

Perhaps those of us who have should start a support group: Dysfunctionary Survivors Anonymous (DSA)?

Right about now, readers may be asking themselves: "If a manager consistently displays any of the shortcomings listed in the dysfunctional dozen, why not replace the manager? Why isn't such a loser fired?"

These are, of course, logical questions, and they identify the proper solution. However, the solutions are not easy to effect. Three major problems stand in the way.

The first problem is the manager. Keep in mind that dysfunctional managers are dysfunctional—but not necessarily stupid. In fact, because these managers are dysfunctional, they have learned through experience (as a tenet to their own self-preservation) to hide their incompetence and disability from those who have the power to take proper corrective action. They cover their dysfuntionalism with ambition. Ambition? Yes, but remember what Oscar Wilde said about ambition: "Ambition is the last refuge of the failure."

The second problem has to do with connections. Some dysfunctional managers owe their management positions to connections with those who own the company. The owners may or may not be aware of the dysfunctional manager's inherent problems and may not care. Their friendship or association priorities may rank above (believe it or not!) their concern over lower-caste employees' welfare, well-being, or mental health. Even if known, the owner may choose not to replace even the more truly and completely dysfunctional manager. If the department is making a profit, upper management falsely believes that nothing is wrong. This, of course, is the drawback to the "If it ain't broke, don't fix it" rule of thumb. The dysfunctional manager/employee relationship *is* "broke." It just isn't broke financially. Besides, keep in mind that a particular dysfunctional manager may be working for another higher-up dysfunctional manager. You know what this means: Birds of a feather tend to roost together.

The third problem has to do with the organization's culture. For example, some organizations have incorporated Total Quality Management (TQM)—sometimes referred to as Managing Total Quality (MTQ)—into their organizations. TQM is often implemented because an organization experiences a crisis. However, TQM is more often implemented at the instigation of some bored-to-tears top manager who read or heard about it and thinks it sounds like a good idea—an idea that will lead to his or her winning the Baldridge National Quality Award, or at the very least, getting high approval ratings from his higher-ups (the brown-noser syndrome).

I hold the view that TQM is a scam. The author—and this text—supports the view that when employees are empowered (when employees closest to the work are empowered to correct problems or defects on their own) what often takes place is that the middle manager is taken out of the management scheme—the manager becomes redundant and not necessary.

To the dysfunctional manager, TQM provides a shield, a facade that he or she can hide behind. What I am saying here is that the dysfunctional manager is "sometimes" a very cunning, adroit, sly, crafty, and well-connected person who has the uncanny ability to survive; TQM aids this process.

TQM is just another management fad, a so-called panacea for all the ills that plague management ... dreamed up by management theorists and gurus—who have never run anything in their lives—but only thought about it from the half-baked point of view of human resource management.

Let's face it, when you empower employees to make their own decisions, why do you need managers? TQM is not "the" silver bullet—that effective management (i.e., leadership) is.

In order to aid the majority of dysfunctional managers who will read this text (with wonder and surprise—and guilt, if they have the intelligence and the nerve to perform self-examination) and who have difficulty understanding simple ideas and concepts, the term "dysfunctional manager" can be replaced with the term "dysfunctionary."

I know what you might be thinking: Dysfunctionary? Dysfunctionary is not a word. But it doesn't matter whether this word can be found in your standard college dictionary—or even the *Oxford English Dictionary*. The words we all use every day to describe our own personal relationships with dysfunctionaries are most commonly found in dictionaries of slang—or are simply not suitable to be printed here (or anywhere else?). A rose by any other name would smell as sweet, and a skunk still stinks even when called a polecat. Whether I call these burdens to the working world dysfunctional managers or dysfunctionaries, you'll know what—and who—I mean—that is, unless you are totally dysfunctional.

REFERENCES AND RECOMMENDATIONS

Angele, F.J., Sr. 1974. *Cross Connections and Backflow Protection*. 2nd ed. Denver, CO: American Water Association.

Capra, F. 1982. *The Turning Point: Science, Society and the Rising Culture*. New York: Simon & Schuster, p. 30.

Daly, H.E. 1980. Introduction to the steady-state economic. In: *Ecology, Ethics: Essays Toward a Steady State Economy*. New York: W.H. Freeman & Company.

De Villiers, M. 2000. *Water: The Fate of Our Most Precious Resource*. Boston, MA: Mariner Books.

Drinan, J.E. 2001. *Water & Wastewater Treatment: A Guide for the Non-engineering Professional*. Boca Raton, FL: CRC Press.

EPA. 2003. EFAB Newsletter volume 3, issue 2: Providing advice on how to pay for environmental protection: *Diamonds and water*. Accessed 09/27/07 @ www.epa.gov/efinpage/efab/newslaters/newsletters6.htm.

Garcia, M.L. 2001. *The Design and Evaluation of Physical Protection Systems*. Butterworth-Heinemann, Oxford, UK.

Gleick, P.H. 1998. *The World's Water 1998-1999: The Biennial Report on Freshwater Resources*. Washington, DC: Island Press.

Gleick, P.H. 2000. *The World's Water 2000-2001: The Biennial Report on Freshwater Resources*. Washington, DC: Island Press.

Gleick, P.H. 2004. *The World's Water 2004-2005: The Biennial Report on Freshwater Resources*. Washington, DC: Island Press.

Harper, S. 2007. Grants to Fuel Green Research. *The Virginian-Pilot*, Norfolk, VA, June 30.

Holyningen-Huene, P. 1993. *Reconstructing Scientific Revolutions*. Chicago, IL: University of Chicago, p. 134.

IBWA. 2004. *Bottled Water Safety and Security*. Alexandria, VA: International Bottled Water Association.

Johnson, R., & Moore, A. 2002. *Opening the Floodgates: Why Water Privatization Will Continue*. Policy Brief 17. Reason Public Institute. [www.rppi.org.pbrief17].

Jones, B.D. 1980. *Service Delivery in the City: Citizens Demand and Bureaucratic Rules*. New York: Longman, p. 2.

Jones, F.E. 1992. *Evaporation of Water*. Chelsea, MI: Lewis Publishers.

Lewis, S.A. 1996. *The Sierra Club Guide to Safe Drinking Water*. San Francisco, CA: Sierra Club Books.

Mather, J.R. 1984. *Water Resources: Distribution, Use, and Management*. New York: John Wiley & Sons.

McGhee, T.J. 1991. *Water Supply and Sewerage*. 6th ed. New York: McGraw-Hill, Inc.

Meyer, W.B. 1996. *Human Impact on Earth*. New York: Cambridge University Press.

Peavy, H.S., et al. 1985. *Environmental Engineering*. New York: McGraw-Hill, Inc.

Pielou, E.C. 1998. *Fresh Water*. Chicago, IL: University of Chicago Press.

Powell, J.W. 1904. *Twenty-Second Annual Report of the Bureau of American Ethnology to the Secretary of the Smithsonian Institution, 1900-1901.* Washington, DC: Government Printing Office.

Spellman, F.R. 2013. *Handbook of Water and Wastewater Treatment Plant Operations.* 3rd ed. Boca Raton, FL: Lewis Publishers.

Turk, J., & Turk, A. 1988. *Environmental Science.* 4th ed. Philadelphia, PA: Saunders College Publishing.

USEPA. 2006. *Watersheds.* Accessed 12/06/19 @ http://epa.gov/owow/watershed/whatis.html.

USEPA. 2005. *Water and Wastewater Security Product Guide.* Accessed 6/06/19 @ http://cfpub.epa.gov.safewater/watersecurity/guide.

U.S. Fish and Wildlife. 2007. *Nutrient Pollution.* Accessed 09/26/07 @ http://fws.gov/chesapeakebay/nutrient.htm.

USGS. 2004. *Estimated Use of Water in the United States in 2000.* Washington, DC: U.S. Geological Survey.

USGS. 2006. *Water Science in Schools.* Washington, DC: U.S. Geological Survey.

U.S. News Online. 2000. *USGS Says Water Supply Will Be One of Challenges in Coming Century.* Accessed 09/20/19 @ http://uswaternews.com/archives/arcsupply/tusgay3.html.

3 Water/Wastewater Infrastructure
Energy Efficiency and Sustainability

INTRODUCTION

The US economy, because it's so energy wasteful, is much less efficient than either the European or Japanese economies. It takes twice as much energy to produce a unit of GDP as it does in Europe and Japan. So, we're fundamentally less efficient and therefore less competitive, and the sooner we begin to tighten up, the better it will be for our economy and society.

Hazel Henderson, on *ENN Radio*

There are a number of long-term economic, social, and environmental trends [Elkington's (1999) so-called Triple Bottom Line] evolving around us. Many of these long-term trends are developing because of us and specifically for us, or simply to sustain us. Many of these long-term trends follow general courses and can be described by the jargon of the day; that is, they can be alluded to or specified by a specific buzzword or buzzwords common in usage today. We frequently hear these buzzwords used in general conversation (especially in abbreviated texting form). Buzzwords such as *empowerment*, *outside the box*, *streamline*, *wellness*, *synergy*, *generation X*, *face time*, *exit strategy*, *clear goal*, and so on and so forth are just part of our daily vernacular.

In this book, the popular buzzword we are concerned with, *sustainability*, is often used in business. However, in water and wastewater treatment, sustainability is much more than just a buzzword; it is a way of life (or should be). There are numerous definitions of sustainability that are overwhelming, vague, and/or indistinct. For our purposes, there is a long definition and a short definition of sustainability. The long definition: ensuring that water and wastewater treatment operations occur indefinitely without negative impact. The short definition: the capacity of water and wastewater operations to endure. Whether we define in a long or short fashion, what does sustainability really mean in the real world of water and wastewater treatment operations?

We defined sustainability in what we call long and short terms. Note, however, that sustainability in water and wastewater treatment operations can be characterized in broader or all-encompassing terms than those simple definitions. As mentioned, using the Triple Bottom Line scenario, in regard to sustainability, the environmental, economic, and social aspects of water and wastewater treatment operations can define today's and tomorrow's needs more specifically.

Infrastructure is another term used in this text; it can be used to describe water and wastewater operations on the whole or can identify several individual or separate elements of water and wastewater treatment operations. For example, in wastewater operations we can devote extensive coverage to wastewater collection and interceptor systems, lift or pumping stations, influent screening, grit removal, primary clarification, aeration, secondary clarification, disinfection, outfalling, and a whole range of solids handling unit processes. On an individual basis each of these unit processes can be described as an integral infrastructure component of the process. Or, holistically, we simply could group all unit processes as one, as a whole, combining all wastewater treatment plant unit processes as "the" operational infrastructure. We could do the same for water treatment operations. For example, as individual water treatment infrastructure components, fundamental systems, or unit processes, we could also list source water intake, pretreatment, screening, coagulation and mixing, flocculation, settling and biosolids processing, filtering, disinfection, storage, and distribution systems. Otherwise we could simply describe water treatment plant operations as the infrastructure.

How one chooses to define infrastructure is not important. What is important is to maintain and manage infrastructure in the most efficient and economical manner possible to ensure its sustainability. This is no easy task. Consider, for example, the 2009 Report Card (which has not changed in 2019) for American Infrastructure produced by American Society of Civil Engineers shown in Table 3.1.

Not only must water and wastewater treatment managers maintain and operate aging and often underfunded infrastructure, but they must also comply with

TABLE 3.1
2009 Report Card for American Infrastructure

Infrastructure	Grade
Bridges	C
Dams	D
Drinking water	D–
Energy	D+
Hazardous waste	D
Rail	C–
Roads	D–
Schools	D
Wastewater	D–
America's infrastructure GPA: **D**	

Source: Modified from American Society of Civil Engineers (2012). *Report Card for American Infrastructure 2009.* Accessed 01/04/2019 @ http://infrastrucutrereportcard.org/.

stringent environmental regulations and keep stakeholders and ratepayers satisfied with operations and with rates. Moreover, in line with these considerations, managers must incorporate economic considerations into every decision. For example, as mentioned, they must meet regulatory standards for the quality of treated drinking water and outfalled wastewater effluent. They must also plan for future upgrades or retrofits that will enable the water or wastewater facility to meet future water quality and future effluent regulatory standards. Finally, and most importantly, managers must optimize the use of manpower, chemicals, and electricity.

SUSTAINABLE WATER/WASTEWATER INFRASTRUCTURE

EPA (2012) points out that *sustainable development* can be defined as that which meets the needs of the present generation without compromising the ability of future generations to meet their needs. The current U.S. population benefits from the investments that were made over the past several decades to build our nation's water/wastewater infrastructure.

Practices that encourage water and wastewater sector utilities and their customers to address existing needs so that future generations will not be left to address the approaching wave of needs resulting from aging water and wastewater infrastructure must continuously be promoted by sector professionals. To be on a sustainable path, investments need to result in efficient infrastructure and infrastructure systems and be at a pace and level that allow the water and wastewater sectors to provide the desired levels of service over the long term.

Sounds easy enough: the water/wastewater manager simply needs to put his or her operation on a sustainable path; moreover, he or she can simply accomplish this by investing. Right? Well, investing what? Investing in what? Investing how much? These are questions that require answers, obviously. Before moving on with this discussion, it is important first to discuss plant infrastructure basics (focusing primarily on wastewater infrastructure and in particular on piping systems) and second to discuss funding (the cash cow vs. cash dog syndrome).

Note that water and wastewater treatment plants typically have a useful life of 20–50 years before they require expansion or rehabilitation. Collection, interceptor, and distribution pipes have life cycles that can range from 15 years to 100 years, depending on the type of material used and where the pipes are laid. Long-term corrosion reduces a pipe's carrying capacity, requiring increasing investments in power and pumping. When water or wastewater pipes age to that point of failure, the result can be contamination of drinking water, the release of wastewater into surface waters or our basements, and high costs of both replacing the pipes and repairing any resulting damage. With pipes, the material used and proper installation of the pipe can be a greater determiner of failure than age.

CASH COWS OR CASH DOGS?

Maintaining the sustainable operations of water and wastewater treatment facilities is expensive. If funding is not available from federal, state, or local governmental entities, then the facilities must be funded by ratepayers. Water and wastewater treatment plants are usually owned, operated, and managed by the community (the municipality) where they are located. While many of these facilities are privately owned, the majority of water treatment plants (WTPs) and wastewater treatment plants (WWTPs) are publicly owned treatment works (POTW; i.e., owned by local government agencies).

These publicly owned facilities are managed and operated on site by professionals in the field. On-site management, however, is usually controlled by a board of elected, appointed, or hired directors/commissioners, who set policy, determine budget, plan for expansion or upgrading, hold decision-making power for large purchases, set rates for ratepayers, and in general control the overall direction of the operation.

When final decisions on matters that affect plant performance are in the hands of, for example, a board of directors comprising elected and appointed city officials, their knowledge of science and engineering, as well as their hands-on on-site problem-solving –experience, can range from everything to nothing. Matters that are of critical importance to those in on-site management may mean little to those on the board. The board of directors may indeed also be responsible for other city services and have an agenda that encompasses more than just the water or wastewater facility. Thus, decisions that affect on-site management can be affected by political and financial concerns that have little to do with the successful operation of a WTP or POTW.

Finances and funding are always of concern, no matter how small or large, well supported or underfunded the municipality. Publicly owned treatment works are generally funded from a combination of sources. These include local taxes, state and federal monies (including grants and matching funds for upgrades), as well as usage fees for water and wastewater customers. In smaller communities, in fact, the water/wastewater (W/WW) plants may be the only city services that actually generate income. This is especially true in water treatment and delivery, which is commonly looked upon as the cash cow of city services. As a cash cow, water treatment works generate cash in excess of the amount of cash needed to maintain the treatment works. These treatment works are "milked" continuously with as little investment as possible. Funds generated by the facility do not always stay with the facility. Funds can be reassigned to support other city services—and when the time comes for a facility upgrade, funding for renovations can be problematic. On the other end of the spectrum, spent water (wastewater) treated in a POTW is often looked upon as one of the cash dogs of city services. Typically, these units make only enough money to sustain operations. This is the case, of course, because managers and oversight boards or commissions are fearful, for political reasons, of charging ratepayers too much for treatment services. Some progress has been made, however, in marketing and selling treated wastewater for reuse in industrial cooling applications and some irrigation projects. Moreover, wastewater solids have been reused as soil amendments; also, ash from incinerated biosolids has been used as a major ingredient in forming cement revetment blocks used in areas susceptible to heavy erosion from river and sea inlets and outlets (Drinan & Spellman 2012).

> **DID YOU KNOW?**
>
> More than 50 percent of Americans drink bottled water occasionally or as their major source of drinking water—an astounding fact given the high quality and low cost of tap water in the United States.

Planning is essential for funding, for controlling expenses, and for ensuring water and wastewater infrastructure sustainability. The infrastructure we build today will be with us for a long time and, therefore, must be efficient to operate, offer the best solution in meeting the needs of a community, and be coordinated with infrastructure investments in other sectors such as transportation and housing. It is both important and challenging to ensure that a plan is in place to renew and replace the infrastructure at the right time, which may be years away. Replacing an infrastructure asset too soon means not benefiting from the remaining useful life of that asset. Replacing an asset too late can lead to expensive emergency repairs that are significantly more expensive than those which are planned (EPA 2012). Additionally, making retrofits to newly constructed infrastructure that was not designed or constructed correctly is expensive. Doing the job correctly the first time requires planning and a certain amount of competence.

THE WATER/WASTEWATER INFRASTRUCTURE GAP

A 2002 EPA report referenced a Water Infrastructure Gap Analysis that compared current spending trends at the nation's drinking water and wastewater treatment facilities to the expenses they can expect to incur for both capital and operations and maintenance costs. The "gap" is the difference between the projected and the needed spending and was found to be over $500 billion over a 20-year period. This important 2002 EPA gap analysis study is just as pertinent today as it was 18 years ago. Moreover, the author draws upon many of the tenets presented in the EPA analysis in formulating many of the basic points and ideas presented herein.

ENERGY EFFICIENCY: WATER/WASTEWATER TREATMENT

Public utility managers are or should be concerned about water and wastewater infrastructure. This could mean they are concerned with the pipes, treatment plants, and other critical components that deliver safe drinking water to our taps and remove wastewater (sewage) from our homes and other buildings and clean it and outfall it to a water body nearby. Although any component or system that makes up water and wastewater infrastructure is important, remember that no water-related infrastructure can function without the aid of some motive force. This motive force (energy source) can be provided by gravitational pull, mechanical means, or electrical energy. We simply can't sustain the operation of water and wastewater infrastructure without energy. As a case in point, consider that drinking water and wastewater systems account for approximately 3–4 percent of energy use in the United States, adding over 45 million tons of greenhouse gases annually. Further, drinking water and wastewater plants are typically the largest energy consumers of municipal governments, accounting for 30–40 percent of total energy consumed. Energy as a percent of operating costs for drinking water systems can also reach as high as 40 percent and is expected to increase 20 percent in the next 15 years due to population growth and tightening drinking water regulations (Spellman 2013).

Not all the news is bad, however. Studies estimate potential savings of 15–30 percent that are "readily achievable" in water and wastewater treatment plants, with substantial financial returns in the thousands of dollars and with payback periods of only a few months to a few years.

In the chapters that follow, we begin our discussion of energy efficiency for sustainable infrastructure in water and wastewater treatment plant operations with brief characterizations of the water and wastewater treatment industries. Later we follow this with a discussion on how public utility managers determine energy usage and how to cut energy usage and costs.

REFERENCES AND RECOMMENDED READING

Drinan, J.E., & Spellman, F.R. 2012. *Water & Wastewater Treatment: A Guide for the Nonengineering Professional.* 2nd ed. Boca Raton, FL: CRC Press.

Elkington, J. 1999. *Cannibals with Forks.* New York: Wiley.

EPA. 2002. *The Clean Water and Drinking Water Infrastructure Gap Analysis.* Washington, DC: United States Environmental Protection Agency—EPA-816-R-02-020.

EPA. 2012. *Frequently Asked Questions: Water Infrastructure and Sustainability.* Accessed 04/01/2019 @ http://water.ep.gov/infrastrure/sustain/si_faqs.cfm.

Spellman, F.R. 2013. *Handbook of Water and Wastewater Treatment Plant Operations.* 3rd ed. Boca Raton, FL: CRC Press.

4 Characteristics of Wastewater and Drinking Water Industries

INTRODUCTION (USEPA 2002)

In this chapter a discussion of the characteristics of the wastewater and drinking water industries provides a useful context for understanding the differences between the industries and how these differences necessitate the use of different methods in estimating needs and costs, and in instituting energy efficiency procedures to ensure sustainability.

WASTEWATER TREATMENT

According to the Code of Federal Regulations (CFR) 40 CFR Part 403, regulations were established in the late 1970s and early 1980s to help publicly owned treatment works (POTW) control industrial discharges to sewers. These regulations were designed to prevent pass-through and interference at the treatment plants and interference in the collection and transmission systems.

Pass-through occurs when pollutants literally "pass through" a POTW without being properly treated, and cause the POTW to have an effluent violation or increase the magnitude or duration of a violation.

Interference occurs when a pollutant discharge causes a POTW to violate its permit by inhibiting or disrupting treatment processes, treatment operations, or processes related to sludge use or disposal.

DRINKING WATER TREATMENT

Municipal water treatment operations and associated treatment unit processes are designed to provide reliable, high quality water service for customers, and to preserve and protect the environment for future generations.

Water management officials and treatment plant operators are tasked with exercising responsible financial management, ensuring fair rates and charges, providing responsive customer service, providing a consistent supply of safe potable water for consumption by the user, and promoting environmental responsibility.

The Honeymoon Is Over

The modern public water supply industry has come into being over the course of the last century. From the period known as the "Great Sanitary Awakening," that eliminated waterborne epidemics of diseases such as cholera and typhoid fever at the turn of the last century, we have built elaborate utility enterprises consisting of vast pipe networks and amazing high-tech treatment systems. Virtually all of this progress has been financed through local revenues. But in all this time, there has seldom been a need to provide for more than modest amounts of pipe replacement, because the pipes last so very long. We have been on an extended honeymoon made possible by the long life of the pipes and the fact that our water systems are relatively young. Now the honeymoon is over. (AWWA 2001)

CHARACTERISTICS OF WASTEWATER INDUSTRY

Wastewater treatment takes effluent (spent water) from water users (consumers, whether from private homes, business, or industrial sources) as influent to wastewater treatment facilities. The wastestream is treated in a series of steps (unit processes, some similar to those used in treating raw water and others that are more involved), then discharged (outfalled) to a receiving body, usually a river or stream.

In the United States, there are 16,024 publicly owned treatment works for treating municipal wastewater. Although there are also some privately owned wastewater treatment works, most of the industry (98 percent) is in fact municipally owned. These POTWs provide service to 190 million people, representing 73 percent of the total population [at the time of the 1996 Clean Water Needs Survey Report to Congress (verified in 2008)]. Seventy-one percent of the facilities serve populations of less than 10,000 people. Furthermore, approximately 25 percent of households in the nation are not connected to centralized treatment, instead using on-site systems (e.g., septic tanks). Although many of these systems are aging or improperly functioning, this text is restricted to centralized collection and treatment systems.

WASTEWATER TREATMENT PROCESS: THE MODEL

Figure 4.1 shows a basic schematic of a centralized conventional wastewater treatment process providing primary and secondary treatment using the *activated sludge process*. This is the model, the prototype, the paradigm used in this book. Though it is true that in secondary treatment (which provides biochemical oxygen demand [BOD] removal beyond what is achievable by simple sedimentation) there are actually three commonly used approaches (**trickling filter**, **activated sludge**, and **oxidation ponds**), we focus, for instructive and illustrative purposes, on the activated sludge process throughout this book. The purpose of Figure 4.1 is to allow the reader to follow the treatment process step by step as it is mentioned and to assist in demonstrating how all the various unit processes sequentially follow and tie into each other.

Wastewater and Drinking Water Industries

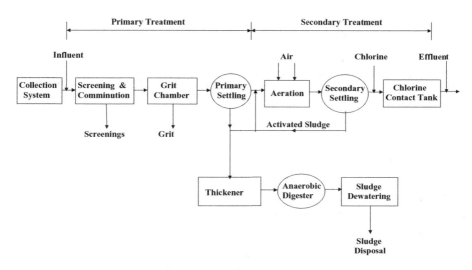

FIGURE 4.1 Unit processes for wastewater treatment.

DID YOU KNOW?

EPA estimates that nationwide capital investment needs for wastewater pollution control is $298.1 billion. This figure represents documented needs for up to a 20-year period. The estimate includes $193.3 billion for wastewater treatment and collection systems, $63.6 billion for combined sewer, and $42.3 billion for stormwater management (EPA 2010).

CHARACTERISTICS OF THE DRINKING WATER INDUSTRY

Water treatment brings raw water up to drinking water quality. The process this entails depends on the quality of the water source. Surface water sources (lakes, rivers, reservoirs, and impoundments) generally require higher levels of treatment than groundwater sources. Groundwater sources may incur higher operating costs from machinery but may require only simple disinfection (see Figure 4.2).

In this text, we define water treatment as any unit process that changes/alters the chemical, physical, and/or bacteriological quality of water with the purpose of making it safe for human consumption and/or appealing to the customer. Treatment also is used to protect the water distribution system components from corrosion. A summary of basic water treatment processes (many of which are discussed in this chapter) is presented in Table 4.1.

The drinking water industry has over 10 times the number of systems as the wastewater industry. Of the almost 170,000 public water systems, 54,000 systems are community water systems that collectively serve more than 264 million people.

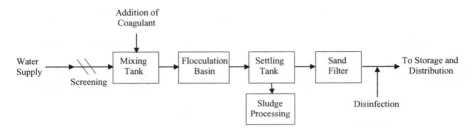

FIGURE 4.2 Unit processes for water treatment.

TABLE 4.1
Basic Water Treatment Processes

Process/Step	Purpose
Screening	Removes large debris (leaves, sticks, fish, etc.) that can foul or damage plant equipment
Chemical pretreatment	Conditions the water for removal of algae and other aquatic nuisances
Presedimentation	Removes gravel, sand, silt, and other gritty materials
Microstraining	Removes algae, aquatic plants, and small debris
Chemical feed and rapid mix	Adds chemicals (coagulants, pH, adjusters, etc.) to water
Coagulation/flocculation	Converts nonsettleable or settable particles
Sedimentation	Removes settleable particles
Softening	Removes hardness-causing chemicals from water
Filtration	Removes particles of solid matter which can cause biological contamination and turbidity
Disinfection	Kills disease-causing organisms
Adsorption using granular activated carbon	Removes radon and many organic chemicals such as pesticides, solvents, and trihalomethanes
Aeration	Removes volatile organic chemicals (VOCs), radon H_2S, and other dissolved gases; oxidizes iron and manganese
Corrosion control	Prevents scaling and corrosion
Reverse osmosis, electrodialysis	Removes nearly all inorganic contaminants
Ion exchange	Removes some inorganic contaminants, including hardness-causing chemicals
Activated alumina	Removes some inorganic contamination
Oxidation filtration	Removes some inorganic contaminants (e.g., iron, manganese, and radium)

Source: Adapted from F.R. Spellman 2008. *Handbook of Water and Wastewater Treatment Plant Operations,* 2nd ed. Boca Raton, FL: CRC Press.

A community water system serves more than 25 people a day all year round. The remaining 114,000 water systems are transient noncommunity water systems (e.g., camp grounds) or non-transient noncommunity water systems (e.g., schools). The scope of discussion in this text is largely confined to community water systems, as these systems serve most of the population. Small systems serving fewer than 10,000 people comprise 93 percent of all community water systems in the nation. However, most of the population (81 percent) receives drinking water from larger systems.

DID YOU KNOW?

In contrast to the wastewater industry, only about 43 percent of community drinking water systems are publicly owned. Most of these systems are under the authority of local governments. Ownership type varies by system size—with almost 90 percent of systems serving more than 10,000 people under public ownership.

CAPITAL STOCK AND IMPACT ON OPERATIONS/MAINTENANCE[*]

The different components of capital stock (total physical capital) that make up the wastewater and drinking water systems in the United States vary in complexity, materials, and the degree to which they are subjected to wear and tear. The expenditures that utilities must make to address the maintenance of systems are largely driven by the condition and age of the components of infrastructure.

USEFUL LIFE OF ASSETS

The life of an asset can be estimated based on the material, but many other factors related to environment and maintenance can affect the useful life of a component of infrastructure. It is not feasible to conduct a condition assessment of all wastewater and drinking water infrastructure systems throughout the United States. However, approximation tools can be used to estimate the useful life of these infrastructure systems.

One approximation tool that can be used to estimate is the *useful life matrix*. This matrix can serve as a tool for developing initial cost estimates and for long-range planning and evaluating programmatic scenarios. Table 4.2 shows an example of a matrix developed as an industry guide in Australia. Although the useful life of a component will vary according to the materials, environment, and maintenance, matrices such as that shown in Table 4.2 can be used at the local level as a starting point for repair and replacement, strategic planning, and cost projects. The United States as well as other industrialized countries have engineering and design manuals that inform professional designers on the accepted standards of practice for design

---

[*] From USEPA. 2002. *The Clean Water and Drinking Water Infrastructure Gap Analysis*. Washington, DC.

TABLE 4.2
Useful Life Matrix

Years	Component
	Wastewater
80–100	Collections
50	Treatment Plants—Concrete Structures
15–25	Treatment Plants—Mechanical & Electrical
25	Force Mains
50	Pumping Stations—Concrete Structures
15	Pumping Stations—Mechanical & Electrical
90–100	Interceptors
	Drinking Water
50–80	Reservoirs & Dams
60–70	Treatment Plants—Concrete Structures
15–25	Treatment Plants—Mechanical & Electrical
65–95	Trunk Mains
60–70	Pumping Stations—Concrete Structures
25	Pumping Stations—Mechanical & Electrical
65–95	Distribution

Source: Adapted from the *International Infrastructure Management Manual*, Version 1.0. Australia/New Zealand Edition, 2000.

life considerations. The U.S. Army Corps of Engineers, the American Society for Testing Materials, the Water Environment Federation (WEF), the American Society of Civil Engineers (ASCE), and several associates maintain data that provides guidance on design and construction of conduits, culverts, and pipes and related design procedures.

Most of the assets of both wastewater and drinking water treatment systems comprise pipes. The useful life of pipes varies considerably based on a number of factors. Some of these factors include the material of which the pipe is made, the conditions of the soil in which it is buried, and the characteristics of the water or wastewater flowing through it. In addition, pipes do not deteriorate at a constant rate. During the initial period following installation, the deterioration rate is likely to be slow, and repair and upkeep expenses low. For pipes, this initial period may last several decades. Later in the life cycle, pipes will deteriorate more rapidly.

The best way to determine the remaining useful life of a system is to conduct periodic condition assessments. At the local level, it is essential for local service providers to complete periodic condition assessments in order to make the best life-cycle decisions regarding maintenance and replacement.

OPERATING AND MAINTAINING (O&M) CAPITAL STOCK

Since 1970, spending in constant dollars on operations and maintenance (O&M) for wastewater treatment and drinking water treatment operations has grown significantly. In 1994, for example, 63 percent of the total spending for wastewater operations and 70 percent of the total spending for drinking water operations were on O&M (CBO 1999).

Likely explanations for the increase in wastewater and drinking water O&M costs include the following:

- Expansion and improvement of services, which translated into an increase in capital stock and a related increase in operations and maintenance costs
- Aging infrastructure, which requires increasing repairs and increasing maintenance costs

Additionally, increases in wastewater operations and maintenance have been driven, in large part, by a large number of solids handling (biosolids) facilities coming on-line. The installation of these facilities has increased O&M costs beginning in the mid-1980s.

Over the next 20 years, O&M expenses are likely to increase in response to the aging of the capital stock: that is, as infrastructure begins to deteriorate, the costs of maintaining and operating the equipment will increase. An American Water Works Association (AWWA) study found that expenditures for deteriorating infrastructure would increase steadily over the next 30 years (AWWA 2001). The projected increase in O&M costs finds support in the historical spending data, which indicate an upward trend for O&M.

Increasing O&M needs will present a significant challenge to the financial resources of wastewater and drinking water systems. As the nation's water infrastructure ages, systems should expect to spend more on O&M. Some systems might even postpone capital investments to meet the rising costs of O&M—assuming that their total level of spending remains constant. The majority of systems likely would increase spending to ensure that both capital and O&M needs are fulfilled, and thus total spending would increase significantly. Many systems would recognize that delaying new capital investments would only increase expenditures on O&M, as old and deteriorated infrastructure would need to be maintained at increasingly higher costs.

DID YOU KNOW?

Pipes are expensive, but invisible. Most people do not realize the huge magnitude of the capital investment that has been made to develop the vast network of distribution mains and pipes—the infrastructure—that makes clean and safe water available at the turn of a tap. Water is by far the most capital intensive of all utility services, mostly due to the cost of pipes, which comprise the water infrastructure that is literally a buried treasure beneath our streets (AWWA 2001).

WASTEWATER CAPITAL STOCK

As mentioned, the basic components of wastewater treatment infrastructure are collection-interceptor systems and treatment works. Systems vary across the clean water industry as a function of the demographic and topographic characteristics of the service area, the unique characteristics of the particular wastestream, and the operating requirements indicated in the permit conditions. The type of treatment is largely controlled by discharge limitations and performance specified through state or federal permits.

Pipe networks represent the primary component of a wastewater treatment system. During the last century, as population grew and spread out from urban centers, the amount of pipes used increased as homes were connected to centralized treatment. Although there is not an actual inventory of the total amount of sewer pipes associated with wastewater collection systems in the United States, the American Society of Civil Engineers has developed an estimate based on feet of sewer per capita—with the average length estimated at 21 feet of sewer per capita. The range varied from 18 feet to 23 feet per capita. The resulting estimate is about 600,000 miles of publicly owned pipe (ASCE 1999).

Because there is no nationwide inventory of wastewater collection systems, the actual age of sewer pipes is not known. However, it is safe to say that installation of pipes has followed demographic increases in population and growth in metropolitan areas associated with suburbanization.

The vast majority of the pipe network in the United States was installed after the Second World War, and the first part of this wave of pipe installation is now reaching the end of its useful life. For this reason, even if the pipe system is extended to accommodate growth and the country invests in the replacement of all pipes that have come to the end of their useful lives, the average age of pipe in the system will still increase until at least 2050.

Although there will be differences based on pipe material and condition, the need to replace pipes will generally echo the original installation wave. Based on the deterioration projections over the next 20 years, if the pipe system is extended to serve growth but there is no renewal or replacement of the existing systems, the amount of pipes classified as "poor," "very poor," or "life elapsed" will increase from 10 percent of the total network to 44 percent of the total network.

Many of the wastewater treatment plants in the United States were completely renovated with major plant expansion and upgrade work beginning in the 1970s, responding to the new treatment requirements of the 1972 Clean Water Act (CWA) and financed to a great extent by EPA's Construction Grants program. Although plants have shorter useful lives than sewer pipes, plant replacement needs are not projected to be a major part of the renewal and replacement requirements until after 2020.

Because plant equipment (e.g., mechanical and electrical components) are easier to observe (i.e., not buried underground—out of sight, out of mind) they are subject to more frequent inspection. Accordingly, some of these visible components will need to be replaced within a 20-year time frame, but relative to the collection systems, they are much less significant.

However, there are implications to the costs associated with the treatment plants. As the treatment plants continue to age, their operation and maintenance costs will increase at a more rapid rate, having a major impact on future operating budgets.

Furthermore, because so many treatment plants were constructed near the same point in time (i.e., beginning in the 1970s), replacement needs will hit at relatively the same time. The initial treatment plant replacement needs will occur at the same time that many pipes installed post-Second World War will begin requiring replacement. Deferral of timely renewal and replacement associated with the oldest pipe over the next 20 years will likely put a system in a difficult financial condition. The typical system could experience a very significant bump in expenditures over a very short period of time to accommodate replacement of old pipes, new pipes, and plant structures in the same time frame.

DRINKING WATER CAPITAL STOCK

The capital stock of an individual drinking water system can be broken down into four principal components: source, treatment, storage, and transmission and distribution mains. Each of these components fulfills an important function in delivering safe drinking water to the public.

While there is no study available that directly addresses the capital makeup of our nation's drinking water systems, a general picture can be obtained from the 2007 EPA Drinking Water Infrastructure Needs Survey. The survey found that the total nationwide infrastructure need is $334.8 billion for the 20-year period from January 2007 to December 2026. Although it is the least visible component of a public water system, the buried pipes of a transmission and distribution network generally comprise most of a system's capital value. Transmission and distribution needs accounted for 60 percent of the total need reported in the 2007 survey. Treatment facilities that are needed to address contaminants with acute and chronic health effects represented the second largest category—with 22 percent of the total need. Storage projects needed to construct or rehabilitate finished water storage tanks represented 11 percent of the total need. Projects needed to address sources of water accounted for 6 percent. The source category included needs for constructing or rehabilitating surface water intakes, raw water pumping facilities, drilled wells, and spring collectors. Neither the storage nor the source categories considered needs associated with the construction or rehabilitation of raw water reservoirs or dams (USEPA 2007).

The need to replace aging transmission and distribution components is a critical part of any drinking water system's capital improvement plan. A recent AWWA 2001 report, "Dawn of the Replacement Era," surveyed the inventory of pipes and the year in which pipes were installed in 20 cities in an effort to predict when the replacement of the pipes would be needed. While the 20 cities in the sample were not selected at random, the cities likely represent a broad range of systems of various ages and sizes from across the country. More importantly, the study provides the only available data on the age of pipes from a reasonably large number of systems.

Age is one factor that affects the life expectancy of pipes. A simple aging model, therefore, was developed to predict when pipes in these 20 cities would need to be replaced. It was assumed that the pipes installed before 1910 would last an average of 120 years; from 1911 to 1945, an average of 100 years; and after 1945, an average of 75 years. In estimating when the current inventory of pipes will be replaced, the model assumes that the actual life span of a pipe will be distributed normally around its expected average life—that is, pipes expected to last 75 years will last 50 to

100 years, those expected to last 100 years will last from 66 to 133 years, and those expected to last 120 years will last 80 to 160 years (AWWA 2001).

This assumption greatly simplifies reality, as the deterioration rates of pipes will vary considerably as a function not only of age, but also of climatic conditions, pipe material, and soil properties. Pipe of the same material, for example, can last from 15 years to over 200 years depending on the soil characteristics alone. In the absence of data that would allow for the development of a model to estimate pipe life (i.e., accounting for local variability of pipe deterioration), the application of a normal distribution to an average life expectancy may provide a reasonable approximation of replacement rates. This model also does not account for other factors, most notably inadequate capacity, that may have equal or greater importance than deterioration in determining pipe replacement rates.

Applying this simple aging model to the historical inventory of pipes for the 20 cities reveals that most of the projected replacement needs for those cities will occur beyond the 20-year period of the analysis—with peak annual replacement occurring in 2040. This conclusion makes sense considering that most of the nation's drinking water lines were installed after the 1940s. Moreover, we need to remember that pipes are hearty but ultimately mortal (AWWA 2001).

DID YOU KNOW?

On average, the replacement cost value of water mains is about $6,300 per household in today's dollars (2001) in the relatively large utilities studied. If water treatment plants, pumps, etc., are included, the replacement cost value rises to just under $10,000 per household, on average (AWWA 2001).

COSTS OF PROVIDING SERVICE

Although many purveyors of water and wastewater services obtain funds from the federal government to finance the costs of capital improvements, most of the funds that systems use both as capital and for operations and maintenance come from revenues derived from user fees. As utilities look to address future capital needs and increasing O&M costs, they need to increase fees to obtain the funding needed for these activities.

While there is no complete source of national data on how rates have changed through time, the State of Ohio has information that can serve as an example for the purposes of a simple discussion. For more than 15 years, the State has conducted an annual survey of water and sewer rates for communities in the state. Data from communities that reported rates for both 1989 and 1999 reveal that there has been an upward shift in the number of communities paying higher annual fees with time.

User rates that are needed to meet the cost of providing service have the potential to negatively impact those segments of the population with low incomes. Data from the Census Bureau show that between 1980 and 1998, incomes at the lower range (as a percentage share of aggregate income for households) declined or stagnated

(U.S. Census Bureau 2000). If rates increase to fund increasing needs, utilities may be challenged to develop rate structures that will minimize impacts on the less affluent segments of society.

REFERENCES AND RECOMMENDED READING

AWWA. 2001. *Dawn of the Replacement Era: Reinvesting in Drinking Water Infrastructure.* Denver, CO: American Water Works Association.

ASCE. 1999. *Optimization of Collection System Maintenance Frequencies and System Performance.* Reston, VA: American Society of Civil Engineers.

CBO. 1999. *Trends in Public Infrastructure.* Washington, DC: Congressional Budget Office.

U.S. Census Bureau. 2000. *The Changing Shape of the Nation's Income Distribution.* U.S. Census P60-204, Current Population Reports Series. Washington, DC: U.S. Census Bureau.

USEPA. 2002. *The Clean Water and Drinking Water Infrastructure Gap Analysis.* Washington, DC: United States Environmental Protection Agency Document EPA-816-R-02-020.

USEPA. 2007. *EPA's 2007 Drinking Water Infrastructure Needs Survey and Assessment.* Washington, DC: United States Environmental Protection Agency.

USEPA. 2010. *Clean Watersheds Needs Survey 2008 Report to Congress.* Washington, DC: United States Environmental Protection Agency.

5 Planning for a Sustainable Energy Future

[S]ustainability means running the global environment—Earth Inc.—like a corporation: with depreciation, amortization and maintenance accounts. In other words, keeping the asset whole, rather than undermining your natural capital.

Maurice Strong

The sustainability revolution is nothing less than a rethinking and remaking of our role in the natural world.

David W. Orr

WASTEWATER AND DRINKING WATER TREATMENT: ENERGY USAGE

Energy represents the largest controllable cost of providing wastewater or water services to the public. Most facilities were designed and built when energy costs were not a major concern. With large pumps, drives, motors, and other equipment operating 24 hours a day, water and wastewater utilities can be among the largest individual energy users in the community. In a National Association of Clean Water Agencies (NACWA) survey of energy use in a typical wastewater treatment plant, 38 percent of energy use was for in-plant pumping, 26 percent was used in aeration, 25 percent for effluent reuse pumping, and 11 percent for other purposes (USEPA 2008; Jones 2006).

PRESENT AND FUTURE CHALLENGES

Wastewater or water treatment plant managers face unprecedented challenges that include ever increasing:

- Public expectations for holding rates/taxes while maintaining service standards
- Population shifts/increases
- Number and complexity of regulatory requirements
- Maintenance and replacement of aging systems/infrastructure
- Concerns about security and emergency preparedness
- Changing workforce demographics
- Challenges in managing personnel, operations, and budgets

Overlaying all these issues are steadily rising energy costs for utilities. Dealing with these rising costs requires public utility managers to ensure better management of their system's energy consumption and to identify areas for improvement. Again, as mentioned, water and wastewater utility energy consumption is generally the largest single sector of a city's energy bill—on the order of 30–60 percent (EIA 2012).

In reviewing a facility's energy performance (i.e., the management of energy consumption and identification of areas for improvement), utility managers may also identify other areas for operational improvements and cost savings, such as labor, chemicals, maintenance, and disposal costs. Additionally, a thorough assessment of a facility's energy performance may alert managers to other issues. An unexplained increase in energy consumption may be indicative of equipment failure, an obstruction, or some other problem within facility operations.

Given these challenges, it is imperative for water and wastewater treatment facilities to investigate implementing systematic programs to minimize energy usage and cost, without sacrificing performance.

Fast Facts[*]

Drinking Water and Wastewater Utility Energy	Wastewater Utility	Drinking Water Utility
• Water and wastewater industries account for an estimated 75 billion kWh of overall U.S. electricity demand.	• There are 15,000 wastewater systems, including 6,000 publicly owned treatment works (POTWs) in the United States.	• There are 60,000 community drinking systems in the United States.
• Drinking water and wastewater systems in the United States spend about $4 billion a year on energy to pump, treat, deliver, collect, and clean water.	• Majority of energy use: treatment process (aeration) and pumping.	• Majority of energy use: pumping.
• Energy efficiency investments often have outstanding rates of return and can reduce costs at a facility by 5, 10, or 25 percent or more.	• Energy use affected by: population, influent loading effluent quality, process type, size, and age.	• Energy use affected by: water source, quality, storage, elevation, distance, age, and process.
• Loads expected to increase by 20 percent in next 15 years due to increased populations and more stringent regulations.	• Major processes: collection systems (sewers and pumping stations), wastewater treatment (primary, secondary, and/or tertiary/advanced), biosolids processing, disposal, or reuse.	• Major processes: production, treatment (disinfection), and distribution.

---

[*] Based on information from Energy Efficiency and Renewable Energy (EERE). Energy Efficiency, 2012. Accessed 04/10/19 @ www.eere.energy.gov/industry/saveenergynow/partners/results.cfm. U.S. Dept. of Energy.

- Energy costs for water and wastewater can be one-third of a municipality's total energy bill.
- If drinking water and wastewater systems reduce use by just 10 percent through cost-effective investments, collectively they could save approximately $400 million and 5 billion kWh annually.

BENCHMARK IT!

With regard to improving energy efficiency and sustainability in drinking water and wastewater treatment operations, recall that benchmarking is simply defined (in this text) as the process of comparing the energy usage of one drinking water or wastewater treatment operation to other similar operations. Local utilities of similar size and design are excellent points of comparison. Broadening the search, one can find several resources discussing the "typical" energy consumption across the United States for a water or wastewater utility of a particular size and design.

As mentioned earlier it is important to keep in mind that in drinking water and wastewater treatment utilities (and other utilities and industries), benchmarking is often used by management personnel to increase efficiency and ensure sustainability of energy resources, but is also used to ensure their own self-preservation (i.e., to retain their lucrative positions). With self-preservation as their motive, benchmarking is used as a tool to compare operations with best in-class-like facilities or operations to improve performance and to avoid the on-going trend to privatize water, wastewater, and other public operations. Again, based on personal observation, usually the real work of preventing privatization is delegated to the individual managers in charge of each specific operation because they also have a stake in making sure that their relatively secure careers are not affected by privatization. Moreover, these frontline managers are best positioned to make nut-and-bolt decisions that can increase efficiency in all operations and that can conserve and sustain energy supplies. It can be easily seen that working against privatization is in the interests of these "local" managers as well as their workers because their jobs may be at stake.

In this text, of course, the focus is on the use of the benchmarking tool (see Figure 1.5) to improve water and wastewater operations' efficiency and ensuring a sustainable future for wastewater and water treatment utilities. Before the benchmarking tool is used, an Energy Team should be formed and assigned the task of studying how to implement energy-saving strategies and how to ensure sustainability in the long run. Keep in mind that forming a "team" is not a silver bullet—the team is only as good as its leadership and its members. Again, benchmarking is a process of rigorously measuring your performance against best-in-class operations, and using the analysis to meet and exceed the latter; thus, those involved in the benchmarking process should be the best of the best (Spellman 2009).

COLLECTION OF BASELINE DATA AND TRACKING ENERGY USE

Using the five-stage benchmarking procedure detailed in Figure 1.5, your benchmarking team identifies, locates, and assembles baseline data that can help you in determining what is needed to improve your energy performance. Keep in mind that the data you collect will be compared to those of similar operations in the benchmarking process. The point is, it is important to collect data that is comparable, like oranges to oranges, apples to apples, grapes to grapes, and so forth. It does little good, makes no sense, and wastes time and money to collect nomenclature data from equipment, machinery, and operations that are not comparable to your utility data.

The first step is to determine what data you already have available. At a minimum, have one full year of monthly data for consumption of electricity, natural gas, and other fuels—if you can get three years of data, even better. However, if you don't have data going this far back, use what you have or can easily collect. In addition, if you can get the data at daily or hourly intervals, you may be able to identify a wider range of energy opportunities (USEPA 2008).

Here are several data elements to document and track for your utility in order to review energy improvement opportunities.

- **Water and/or wastewater flows** are key to determining your energy performance per gallon treated. For drinking water, the distance of travel and number of pumps are also key factors.
- **Electricity data** includes overall electricity consumption (kWh) as well as peak demand (kW) and load profiles if available.
- **Other energy data** includes purchases of diesel fuel, natural gas, or other energy sources, including renewables.
- **Design specifications** can help you identify how much energy a given process or piece of equipment should be using.
- **Operating schedules** for intermittent processes will help you make sense of your load profile and possibly plan an energy-saving or cost-saving alternative.

Along with making sure that the data you collect is comparable, that is, apples to apples, etc., keep in mind that energy units may vary. If you are comparing apples to apples, are you comparing bushel to bushel, pound to pound, or quantity to quantity. In an energy efficiency and sustainability benchmarking study comparison, for example, captured methane or purchased natural gas may be measured in 100 cu feet (cubic feet) or Metric Million British Thermal Units (MMBTU). Develop a table like Table 5.1 to document and track your data needs (USEPA 2008).

REMEMBER
Keep units consistent!

Planning for a Sustainable Energy Future

TABLE 5.1
Data Needs

Data Need	Units
Wastewater flow	MGD
Electricity consumption	kWh
Peak demand	kW
Methane capture (applies to plants that digest biosolids)	MMBTU
Microturbine generation	kWh
Natural gas consumed	MMBTU
Fuel oil consumed	Gallons
Diesel fuel consumed	Gallons
Design specifications	N/A
Operating schedules	N/A
Grease trap waste collected (future renewable fuel source)	Gallons
Other (based on your operation)	TBD

Consider any other quantities that you'll want to measure. Is there anything you would add to Table 5.1? The chances are you will add quantities; that is why it is labeled Table 5.1, with other renditions to follow. Let's get back to unit selection for your tables. Make sure you select units that your Energy Team is comfortable with and that your data is typically available in. If the data is reported using the wrong units, you may have some conflicting or confusing results.

Keep in mind that units by themselves are not that informative; to be placed in proper context, they need to be associated with an interval of time. Therefore, for the next step simply expand Table 5.1 by adding another column, "Desired Frequency of Data," to form Table 5.2.

TABLE 5.2
Data Needs

Data Need	Units	Desired Frequency of Data
Wastewater flow	MGD	Daily
Electricity consumption	kWh	Hourly if possible, else daily
Peak demand	kW	Monthly
Methane capture	MMBTU	Monthly
Microturbine generation	kWh	Monthly
Natural gas consumed	MMBTU	Monthly
Fuel oil consumed	Gallons	Monthly
Diesel fuel consumed	Gallons	Monthly
Design specifications	N/A	N/A
Operating schedules	N/A	N/A

Remember, while knowing your utility's energy consumption per month is useful, knowing it in kWh per day is better. With hourly consumption data, you develop a "load profile" or a breakdown of your energy demand during the day; if your load profile is relatively flat, or if your energy demand is greater in the off-peak hours (overnight and early morning) than in the peak hours (daytime and early evening), your utility may qualify for special pricing plans from your energy provider.

Typically, water and wastewater treatment operations have a predictable diurnal variation (i.e., fluctuations that occur during each day). Usage is most heavy during the early overnight. Usage is most heavy during the early morning, lags during the afternoon, has a second, less-intensive peak in the early evening, and hits the lowest point overnight. Normally, energy use for water and wastewater treatment operations could be expected to follow the pattern of water flows. However, this effect can be delayed by the travel time from the source, through the collection system, to the plant, or by storage tanks within the distribution system to customers. A larger system will have varying travel times, whereas a smaller system will have lower variability. Moreover, this effect can be totally eliminated if the plant has an equalization tank (USEPA 2008).

If your utility is paying a great deal of money for peak demand charges, you might consider the capital investment of an equalization tank. Demand charges can be significant for wastewater utilities, as they are generally about 25 percent of the utility's electricity bill (WEF 1997).

The next step is to determine how you will collect baseline data. Energy data is recorded by your energy provider (e.g., electric utility, natural gas utility, or heating oil and diesel oil companies). A monthly energy bill contains the total consumption for that month, as well as the peak demand. In some cases, your local utilities will record the demand on every meter for every fifteen-minute interval of the year. Similar data may be available if you have a system at your utility that monitors energy performance. Sources of energy data include the following:

- **Monthly energy bills** vary in detail but all contain the most essential elements.
- **The energy provider** may be able to provide more detailed information.
- **An energy management program** (e.g., Supervisory Control and Data Acquisition—SCADA) automatically tracks energy data, often with submeters to identify the load on individual components. If such a system is in place at your utility, you will have a large and detailed data set on hand.

SIDEBAR 5.1   WHAT IS SCADA?

Public utility managers must know something about SCADA, especially if it is employed in their utility system. Simply, SCADA is a computer-based system that remotely controls processes previously controlled manually. SCADA allows an operator using a central computer to supervise (control and monitor) multiple networked computers at remote locations. Each remote computer can control mechanical processes (pumps, valves, etc.) and collect data from sensors at its remote location. Hence the name "Supervisory Control and Data Acquisition."

The central computer is called the Master Terminal Unit, or MTU. The operator interfaces with the MTU using software called Human Machine Interface, or HMI. The remote computer is called Programmable Logic Controller (PLC) or Remote Terminal Unit (RTU). The RTU activates a relay (or switch) that turns mechanical equipment "on" and "off." The RTU also collects data from sensors.

In the initial stages utilities ran wires, also known as hardwire or land lines, from the central computer (MTU) to the remote computers (RTUs). Since remote locations were found hundreds of miles from the central location, utilities began to use public phone lines and modems, leased telephone company lines, and radio and microwave communication. More recently, they have also begun to use satellite links, Internet, and newly developed wireless technologies.

Since the SCADA systems' sensors provide valuable information, many utilities establish "connections" between their SCADA systems and their business system. This allows utility management and other staff access to valuable statistics, such as water usage. When utilities later connect their systems to the Internet, they will be able to provide stakeholders with water/wastewater statistics on the utility web pages.

SCADA Applications in Water/Wastewater System

As stated above, SCADA systems can be designed to measure a variety of equipment operating conditions and parameters or volumes and flow rates or water quality parameters, and to respond to changes in those parameters either by alerting operators or by modifying system operation through a feedback loop system without having personnel physically visit each process or piece of equipment on a daily basis to check it and/or to ensure that it is functioning properly. SCADA systems can also be used to automate certain functions, so that they can be performed without needing to be initiated by an operator (e.g., injecting chlorine in response to periodic low chlorine levels in a distribution system, or turning on a pump in response to low water levels in a storage tank). As described above, in addition to process equipment, SCADA systems can also integrate specific security alarms and equipment, such as cameras, motion sensors, lights, data from card reading systems, etc., thereby providing a clear picture of what is happening at areas throughout a facility. Finally, SCADA systems also provide constant, real-time data on processes, equipment, location access, etc., enabling the necessary response to be made quickly. This can be extremely useful during emergency conditions, such as when distribution mains break or when potentially disruptive BOD spikes appear in wastewater influent.

Because these systems can monitor multiple processes, equipment, and infrastructure and then provide quick notification of, or response to, problems or upsets, SCADA systems typically serve as the first-line detection method for atypical or abnormal conditions. For example, a SCADA system connected to sensors that measure specific water quality parameters are measured outside of a specific range. A real-time customized operator interface screen could display and control critical systems monitoring parameters.

The system could transmit warning signals back to the operators, such as by initiating a call to a personal pager. This might allow the operators to initiate actions to prevent contamination and disruption of the water supply. Further automation of the system could ensure that the system initiates measures to rectify the problem. Preprogrammed control functions (e.g., shutting a valve, controlling flow, increasing chlorination, or adding other chemicals) can be triggered and operated based on SCADA utility.

Barton Gellman 2002

Other data needs may also have a range of sources. Design specifications for equipment may be in manuals at your utility but you may still need to contact the manufacturers for specific items. In addition to providing raw data, your energy provider can offer you extensive expertise on energy-saving technologies, practices, and programs as well as contractors who can help you implement certain types of improvements (USEPA 2008).

BASELINE AUDIT

The energy audit is an essential step in energy conservation and energy management efforts. Your drinking water or wastewater operation may have had an energy audit or energy program review conducted at some point. If so, find the final report and have your Energy Team review it. How long did the process take? Who participated in it—your team, the electric utility, independent contractors? What measures were suggested to improve energy efficiency? What measures were actually implemented and did they meet expectations? Were there lessons learned from the process that should be applied to future audits? In addition, if your facility's previous energy audit had recommended measures, determine if they are still viable.

In many cases, electrical utilities offer audits as part of their energy conservation programs. Independent energy service companies also provide these services. An outside review from an electric utility or an engineering company can provide useful input but it is important to ensure that any third party is familiar with water and wastewater systems.

Some energy audits focus on specific types of equipment such as lighting, HVAC, or pumps. Others look at the processes used and take a more systematic approach. Audits focused on individual components, as well as in-depth process audits, will include testing equipment. For example, in conducting the baseline energy audit, the Energy Team may compare the nameplate efficiency of a motor or pump to its actual efficiency.

In a process approach, a preliminary walk-through or walk-around audit is often used as a first step to determine if there are likely to be opportunities to save energy. If such opportunities exist, then a detailed process audit is conducted. This may include auditing the performance of the individual components as well as considering how they work together as a whole. Much like an environmental management system's initial assessment that reviews current status of regulatory requirements, training, communication, operating conditions, and current practices and processes, a preliminary energy audit or energy program review will provide your utility with a baseline of what your energy consumption is at that point in time.

Once you have collected your utility's baseline data and tracked monthly and annual energy use, there are two additional steps to completing your energy assessment or baseline energy audit: conduct a field investigation and create an equipment inventory and distribution of demand and energy (USEPA 2008).

FIELD INVESTIGATION

The field investigation is the heart of an energy audit. It will include obtaining information for an equipment inventory, discussing process operations with the individuals responsible for each operation, discussing the impact of specific energy conservation ideas, soliciting ideas from your Energy Team, and identifying the energy profiles of individual system components. The Electric Power Research Institutes (EPRI) recommends evaluating how each process or piece of equipment could otherwise be used. For example, it might be possible for a given system to be replaced or complemented for normal operation by one of lower capacity; to run fewer hours; to run during off-peak hours; to employ a variable speed drive; and/or to be replaced by a newer or more efficient system. Depending on the situation, one or more of these changes might be appropriate.

CREATE EQUIPMENT INVENTORY AND DISTRIBUTION OF DEMAND AND ENERGY

Equipment inventory and distribution of demand and energy is a record of your operation's equipment, equipment names, nameplate horsepower (if applicable), hours of operation per year, measured power consumption, and total kilowatt-hours (kWh) of electrical consumption per year. Other criteria such as age may also be included. In addition, different data may be appropriate for other types of systems such as methane-fired heat and power systems.

You may find that you already have much of this information in your maintenance management system (if applicable). A detailed approach to developing an equipment inventory and identifying the energy demand of each piece of equipment is provided in the 1997 book *Energy Conservation in Wastewater Treatment Facilities: Manual of Practice No. MFD-2, Water Environment Federation*. The basics are presented here, but readers are encouraged to review the WEF (1997) *Manual of Practice* for a more thorough explanation.

Examples of drinking water and/or wastewater treatment operations equipment inventories and the relevant energy data to collect could include the following:

Motors and Related Equipment
- Start at each motor control center (MCC) and itemize each piece of equipment in order as listed on the MCC.

- Itemize all electric meters on MCCs and local control panels.
- Have a qualified electrician check the power draw of each major piece of equipment.

Pumps
- From the equipment manufacturer's literature, determine the pump's power ratio (this may be expressed in kW/mgd).
- Multiply horsepower by 0.746 to obtain kilowatts.
- Compare the manufacturer's data with field-obtained data.

Aeration Equipment
- Power draw of aeration equipment is difficult to estimate and should be measured.
- Measure aspects related to biochemical oxygen demand (BOD) loading, foot-to-microorganism ratio, and oxygen-transfer efficiency (OTE). Note that OTE levels depend on type and condition of aeration equipment. Actual OTE levels are often considerably lower than described in the literature or in manufacturers' materials.

REFERENCES AND RECOMMENDED READING

EIA. 2012. *The Current and Historical Monthly Retail Sales, Revenues and Average Revenue per Kilowatt Hour by State and by Sector.* Accessed 04/09/20 @ http://www.eia.doe.gov/cneaf/Electricity/page/sales_reveune.xis.

Gellman, B. 2002. Cyber-Attacks by Al Qaeda Feared: Terrorists at Threshold of Using Internet as Tool of Bloodshed, Experts Say. *Washington Post,* June 27, p. A01.

Jones, T. 2006. *Water-Wastewater Committee: Program Opportunities in the Municipal Sector: Priorities for 2006,* Presentation to CEE June Program Meeting, Boston, MA, June 2006.

Spellman, F.R. 2009. *Handbook of Water and Wastewater Treatment Plant Operations.* 2nd ed. Boca Raton, FL: CRC Press.

USEPA. 2008. *Ensuring a Sustainable Future: An Energy Management Guidebook for Wastewater and Water Utilities.* Washington, DC: United States Environmental Protection Agency.

WEF. 1997. *Energy Conservation in Wastewater Treatment Facilities.* Manual of Practice No. MFD-2. Alexandria, VA: Water Environment Federation.

6 Energy Efficient Operating Strategies

For it matters not how small the beginning may seem to be:

what is once well done is well done forever.

Henry David Thoreau

INTRODUCTION*

The California Energy Commission (2012) points out that electrical energy consumption at water and wastewater treatment plants is increasing because of more stringent regulations and customer concerns about water quality. As a result, more facility managers are turning to energy management to reduce operating costs. Reducing energy consumption (by managing your electrical load), however, is only part of the equation. Operating strategies to reduce energy usage also include operation and maintenance practices; these operating strategies are discussed in this chapter.

ELECTRICAL LOAD MANAGEMENT

By choosing when and where to use electricity, wastewater and drinking water facilities can often save as much (or more) money as they could by reducing energy consumption. Note that electricity is typically billed in two ways: by the quantity of *energy* used over a period of time, measured in kilowatt-hours, and by *demand*, the rate of flow of energy, measured in kilowatts.

DID YOU KNOW?

Electricity prices generally reflect the costs to build, finance, maintain, manage, and operate power plants and the electricity grid (the complex system of power transmission and distribution lines), and to operate and administer the utilities that supply electricity to consumers. Some utilities are for-profit, and their prices include a return for the owners and shareholders.

---

* Based on information from EPA. 2012. *Water & Energy Efficiency in Water and Wastewater Facilities.* Accessed 04/21/19 @ http://epa.gov/region09/waterinfrastrucutre/technology.html.

Rate Schedules

Electric utilities often structure rates to encourage customers to minimize demand during peak periods, because it is costly to provide generating capacity for use during periods of peak demand. That's why drinking water and wastewater treatment plants should investigate the variety of rate schedules offered by electric utilities. They may achieve substantial savings simply by selecting a rate schedule that better fits their pattern of electricity use.

- **Time-of-Use Rates**—in most areas of the country time-of-use rates, which favor off-peak electrical use, are available. Under the time-of-use rates, energy and demand charges vary during different block periods of the day. For example, energy charges in the summer may be only 5¢ per kilowatt-hour with no demand charge between 9:30 p.m. and 8:30 a.m., but increase to 9¢ per kilowatt-hour with demand charge of $10 per kilowatt between noon and 6:00 p.m. The monthly demand charge is often based on highest 15-minute average demand for the month.
- **Interruptible Rates**—offer users discounts in exchange for a user commitment to reduce demand on request. On the rare occasions when a plant receives such a request, it can run standby power generators.
- **Power Factor Charges**—power factor, also known as reactive power or kVAR, reflects the extent that current and voltage cycle in phase. Low power factor, such as that caused by a partly loaded motor, results in excessive current flow. Many electric utilities charge extra for low power factor because of the cost of providing the extra current.
- **Future Pricing Options**—as the electrical industry is deregulated, many new pricing options will be offered. *Real-time pricing*, where pricing varies continuously based on regional demand, and *block power*, or electricity priced in low-cost, constant-load increments, are only two of the many rate structures that may be available. Facilities that know how and when they use energy and have identified flexible electric loads can select a rate structure that offers the highest economy, while meeting their energy needs.

Energy Demand Management

Energy demand management, also known as demand side management (DSM), is the modification of consumer (utility) demand for energy through various methods. DSM programs consist of the planning, implementing, and monitoring activities of electric utilities that are designed to encourage consumers to modify their level and pattern of electricity usage.

Energy Management Strategies

- **Conduct an Energy Survey**—the first step to an effective energy management program for your facility is to learn how and when each piece of equipment uses energy. Calculate the demand and monthly energy consumption for the largest motors in your plant. Staff may be surprised at the results—a 100 hp (horsepower) motor may cost over $4,500 per month if run continuously.

The rate at which energy is used will vary throughout the day, depending upon factors such as the demand from the distribution system and reservoir and well levels for water systems or influent flows and biological oxygen demand loading for wastewater systems. Plot daily electrical load as a function of time for different plant loading conditions and note which large equipment can be operated off-peak. Examine all available rate schedules to determine which can provide the lowest cost in conjunction with appropriate operational changes.

- **Reduce Peak Demand**—look for opportunities to improve efficiency of equipment that must run during the peak period, such as improving pump efficiency or upgrading a wastewater plant's aeration system. During on-peak periods, avoid using large equipment simultaneously: two 25-kilowatt pumps that run only two hours each day can contribute 50 kilowatts to demand if run at the same time.
- **Shift Load to Off-Peak**—many large loads can be scheduled for off-peak operation. For example, plants can use system storage to ride out periods of highest load rather than operating pumps. Avoid running large intermittent pumps when operating the main pumps.
- **Improve Power Factor**—low power factor is frequently caused by motors that run less than fully loaded. This also wastes energy because motor efficiency drops off below full load. Examine motor systems to determine if the motor should be resized or if a smaller motor can be added to handle lower loads. Power factor can also be corrected by installing a capacitor in parallel with the offending equipment.

ELECTRICAL LOAD MANAGEMENT SUCCESS STORIES[*]

Encina Wastewater Authority

Service Area: 125 square miles
Wastewater System Capacity: 36 MGD (million gallons per day)
Wastewater Treatment Type: Secondary
Secondary Treatment Method: Activated Sludge
Annual System-Wide Purchased Electricity: $174,300 (2.2 million kWh)
Cogeneration Capacity: 1.4 MW (megawatts)
Annual Savings Attributed to Energy Efficient Strategies: $611,000

How does a wastewater agency continue to operate economically and maintain high quality while serving its rapidly growing customer base? Staff at the Encina Wastewater Authority decided energy efficiency was the answer. They set into motion a comprehensive energy management program addressing every aspect of the facility's energy use, from demand control to lighting retrofits. The plant now profits from

---

[*] From California Energy Commission. 2012. *Success Story Encina Wastewater Authority*. Accessed 04/22/2020 @ www.energy.ca.gov/process/pubs/encina.pdf

increased energy efficiency, operational savings, and a staff more attuned to methods of achieving these benefits.

Key Improvements

Staff of Encina Wastewater Authority integrated the following measures:
- *Use cogeneration to produce on-site electricity and thermal energy*—Encina's system consists of three engine generators that run on purchased natural gas. Heat from the generators maintains a constant 96°F for digesters and is used to heat offices and run three absorption chillers that provide cooling. Although the system can produce 1,425 kW, emission restrictions currently allow use of only two generator engines. The facility's Cogeneration Optimization Project will improve the system's efficiency, giving Encina "qualified facility" status from the Federal Energy Regulatory Commission, which will lower costs for natural gas. Upgrades will reduce emissions by converting engines to "lean burn." Encina's cogeneration facility produces 80 percent of on-site power and provides heat to digesters and HVAC applications that would otherwise operate solely on purchased natural gas and electricity. As it currently operates, the system produced about 8 million kWh/year (kilowatt-hours per year). When upgrades are completed, improvements in emissions will permit a third engine to be brought on-line, increasing generation capacity by 50 percent while continuing to meet air quality restrictions.

DID YOU KNOW?

Aeration in wastewater treatment operations is the mixing air and a liquid by one of the following methods: spraying the liquid in the air; diffusing air into the liquid; or agitating the liquid to promote surface adsorption of air.

- *Use fine bubble diffusers for aeration*—because aeration constitutes as much as 50 percent of an activated sludge plant's energy costs, increased efficiency in this area is critical. When expanding their plant, Encina chose fine bubble diffusers over the less-efficient coarse-bubble versions previously used. Additionally, Encina has automated control of dissolved oxygen levels for over 10 years, using probes throughout the aeration basins to monitor and help maintain dissolved oxygen levels. Again, these automated controls maintain dissolved oxygen levels at predetermined set points. Because fine bubble diffusers transfer more dissolved oxygen into the water than coarse versions, less oxygen needs to be introduced, lowering the energy required to drive dissolved oxygen compressors. The fine bubble diffusers are estimated to save about 2,920,000 kWh/year. Some facilities are reluctant to automate dissolved oxygen controls, because they are concerned that fouled probes will hamper reliable readings.

- *Enact demand control strategies*—Encina's energy management program emphasizes off-peak pumping, enabling the facility to profit from lower utility rates. Staff manually shut down select high-demand equipment during on-peak periods. Because monthly billings are based on the highest energy demand in a 15-minute block, strict compliance is essential. Many wastewater agencies use control systems, but the Encina staff have demonstrated that expensive automated controls are not a prerequisite for success.
- *Pump water more efficiently with variable frequency drives.*
- *Upgrade standard motors to energy efficient motors.*

Moulton Niguel Water District

Service Area: 37.5 square miles
Potable Water System Capacity: 48 MGD
Wastewater System Capacity: 17 MGD
Wastewater Treatment Type: Tertiary
Secondary Treatment Method: Activated sludge
Annual System-Wide Purchased Electricity: $1,310,000 (15.5 million kWh)
Annual Savings Attributed to Energy Efficient Strategies: $332,000

For over a decade, automation and instrumentation have helped Southern California's Moulton Niguel Water District supply water and treat wastewater economically and efficiently. Facing a major rise in energy costs, the agency explored other methods to increase energy efficiency. Working closely with Southern California Edison and San Diego Gas & Electric to identify optimal rate schedules and energy efficiency strategies, the district implemented a program in 1992 that has yielded substantial savings in the reservoir-fed branches of their distribution system. Additionally, the district plans to investigate potential improvements to their potable water systems, requiring full-speed pumping to maintain system pressure.

Key Improvements

Moulton Niguel Water District staff implemented changes in the following areas:

- *Install programmable logic controllers to benefit from lower off-peak utility rates*—Moulton Niguel uses automated controls and programmable logic controllers to enable 77 district pumping stations to benefit from lower off-peak utility rates. The controls activate pumps during off-peak hours, bringing reservoirs to satisfactory levels. During peak hours, pumping is halted, allowing reservoir levels to fall. Stations employing this strategy are on "reservoir duty"—meaning the system is pressurized by the static head of the reservoir. All stations previously operated in "closed grid mode," running pumps 24 hours a day to maintain system pressure.

  The programmable logic controllers' sophisticated internal clock and calendar automatically adjusts for seasonal changes—consistently keeping equipment running off-peak. This strategy has decreased pumping costs (saving nearly $320,000 annually), and allowing reservoir levels to fall has improved water quality.

- *Regulate lift station wastewater levels using proportional, integral, and derivative controls to automatically transmit data to a central computer*— Moulton Niguel previously cycled constant-speed wastewater pump drives on and off to distribute wastewater. As a result, drive control was limited; pump motors were subject to starting surges, and the system shutdown left sewage sitting in pipes, producing offensive odors.

    They decided to replace their standard motor drives with variable frequency drives linked to proportional, integral, and derivative controls. This system regulates wastewater levels by sending a signal to variable frequency drives controllers to modulate wastewater flow. The new proportional, integral, and derivative/variable frequency drives system provides a continuous, modulating flow that uses less energy, reduces motor wear and high energy demands from motor starting surges, ensures that sewage does not remain stagnant in pipes, thereby reducing odor problems, and reduces energy costs by about 4 percent.
- *Install variable frequency drives on the wastewater system to control pump speed in coordination with the proportional, integral, and derivative system, reducing costs.*
- *Specify that all motors used in new construction be 95–97 percent efficient; replace standard-efficiency motors with energy efficient motors (ongoing).*

OPERATION AND MAINTENANCE: ENERGY COST-SAVING PROCEDURES[*]

From 2008 to 2010 EPA developed and conducted energy workshops for over 500 water and wastewater utility associates in a handful of western states. The following subsections identify the utilities that participated in the program and provide a summary of their results. Many of the projects required no additional resources outside of existing staff time and minor equipment purchases made within existing expense accounts. Some facilities focused on collecting and using renewable energy (specifically energy generated from the force of water dropping from an elevation while traveling through pipes). Other sites concentrated on reducing energy consumption by increasing energy efficiency. Some utilities reduced energy use during the day when energy costs more and increased energy use during times of the day when energy costs less. At several of the facilities, optimizing operations resulted in significant savings without requiring a large capital outlay. One step each of the 10 facilities profiled below developed and implemented was an Energy Improvement Management Plan. More details on the specific projects can be found in the case studies developed by the utilities and described in the subsections that follow.

---

[*] From EPA. 2011. *U.S. EPA Region 9 Energy Management Initiatives for Public Wastewater and Drinking Water Utilities Facilitating Utilities Toward Sustainable Energy Management.* Washington, DC: U.S. Environmental Protection Agency.

Chandler Municipal Utilities, Arizona (February 21, 2012)

- **Facility Profile**—Chandler Municipal Utilities is located in the City of Chandler, Arizona, within Maricopa County. The Municipal Utilities Department oversees wastewater treatment, reclaimed water, and the drinking water supply for the city. The utility selected its potable water system for the Energy Management Initiative. The Chandler potable water system serves 255,000 customers. The system treats an average of 52 MGD of groundwater and surface water at two treatment plants. The use of groundwater, from 31 wells, requires more energy than surface water from the Salt River Project and Central AZ Project. Additional energy is needed to bring groundwater to the surface for treatment and distribution.
- **Baseline Data**—Chandler spent $2.9 million on electricity last year treating potable water. The annual electricity used at the water treatment plant is 33,880,000kWh. In 2010 the plant's energy consumption generated 22,268 metric tons of carbon dioxide equivalent (MTC0$_2$) of greenhouse gas (GHG) emissions.
- **Energy Improvement Management Plan**—Chandler Municipal Utilities chose to reduce energy consumption by optimizing the potable water system. They did this in two days. First, they revised tank management practices based on hydraulic modeling and master planning to find the best configuration and operating program to reduce groundwater pumping. Second, the utility staff upgraded pumps and revised pressure zones to operate more efficiently under the new operating program. Based on this energy management approach, Chandler's goal was to reduce the number of kilowatt-hours used to produce and distribute one million gallons of potable water by 5 percent from its 2010 levels. Chandler chose to develop a strong team as an area of focus for the Energy Management System.
- **Challenges**—one of the biggest challenges Chandler faced was changing staff attitudes and long-established habits associated with operating a small groundwater-based system—to the practicality of operating a large surface-water-dominated system. A series of small victories led to a staff-driven team approach to system optimization.

The City of Chandler's potable water production and distribution system expanded rapidly to meet the growth of the 1990s and early 2000s. Wells and mains were added so the system was able to meet all its demands. The recent economic slowdown gave Chandler staff the opportunity to analyze the system as a whole, rather than as a collection of separate parts. Complicating factors included a lack of consistent historic design philosophy and evolution of the system from a small groundwater-based utility to a surface-water-dominated system serving a population of 250,000 and several major industrial and commercial customers.

Surface water is the most cost-effective source of water for Chandler, but due to a lack of dedicated transmission infrastructure, staff had a difficult time filling tanks with surface water. Facing budget constraints, staff revisited all aspects of the system. The result was (1) an expanded second pressure zone; (2) consistent hydraulic grade lines for the pressure zones; (3) focused rehabilitation of key facilities; and (4) a new tank management strategy.

During this time the programmable logic controllers the system used were no longer being supported by the manufacturer, so they were replaced by controllers with much better information collection capabilities. The new technology gave the operators much better information and control of the system. Given the new tools, the operators developed and tested new operation strategies which have resulted in a more robust system that produces better-quality water, while using fewer resources.

- **Accomplishments**—Chandler fell a bit short of the 5 percent goal, but was able to achieve a reduction of 4.2 percent. They also produced 4.7 percent more potable water in 2011 than they did in 2010. Had Chandler not reduced the energy necessary to produce and distribute one million gallons they would have used an additional 1,445,000 more kilowatt-hours. Using the average cost per kilowatt-hour Chandler paid for power in 2011, this amounts to an energy savings of almost $130,000 and avoiding the generation of 950 $MTCO_2$ of GHG emissions.

  One aspect of Chandler's potable water system optimization approach involved using a higher percentage of surface water than in previous years. This resulted in substantial savings in water resource costs, and a significant reduction in chlorine use.

  Chandler adopted a team approach to optimization that resulted in a high level of understanding of system dynamics throughout the organization, an involved staff that continually identifies ways to improve the efficiency and operation of the system, improvements in data acquisition and management, and multiple open channels of communication.
  - Annual Energy Savings: 1,445,000 kWh
  - Annual Cost Savings: $130,000
  - Annual GHG Reductions: 996 $MTCO_2$, equal to the removal of 195 passenger vehicles from the road
  - Project Cost: No additional funds required
  - Payback Period: Immediate
- **Future Steps**—staff are continuing to evaluate system performance and seek additional opportunities for optimization. Some lighting has been upgraded; more lighting upgrades are planned for the future. Staff are investigating the feasibility of on-site power generation using solar panels and in-pipe hydraulic power generation.

AIRPORT WATER RECLAMATION FACILITY, PRESCOTT, ARIZONA (FEBRUARY 21, 2012)

- **Facility Profile**—the Airport Water Reclamation Facility (AWRF) is one of the two facilities owned and operated by the City of Prescott, within Yavapai County. Prescott is positioned close to the center of Arizona between Phoenix and Flagstaff, just outside the Prescott National Forest. The original wastewater treatment plant was built in 1978 and received a

Energy Efficient Operating Strategies 85

major facility upgrade in 1999. The next major upgrade began in 2012. The City of Prescott also operates another wastewater treatment plant called Sundog.
- **Baseline Data**—AWRF treats 1.1 million gal of wastewater per day for approximately 18,000 residents. The facility spends $160,000 annually on electricity costs and uses 1.8 million kWh. GHG emissions for the SWTF are 1,023 MTCO$_2$.
- **Energy Improvement Management Plan**—the City of Prescott elected to construct a hydro turbine electric generation unit as part of the Energy Improvement Management Plan at the Airport Water Reclamation Facility to conserve energy and better manage resources. The turbine will be placed at the discharge point of the recharge water pipeline to convert the potential energy in the flowing water to electricity. This hydro turbine has the potential to produce 125,000 kWh/year, which would save the City approximately $12,000 per year.
- **Challenges**—the greatest challenge so far has been the time commitment required to accomplish the project concurrent with the design of the facility expansion. Initially, there was also some difficulty connecting with the appropriate staff at the power company; however, that has since been resolved.
- **Accomplishments**—energy and cost savings will result with the new hydro turbine electric generation unit installation at the Airport Water Reclamation Facility. This project is estimated to produce 125,000 kWh and save $12,000 in electrical costs per year. The energy management program has promoted an awareness of energy uses and potential savings associated with minor and major changes to operations. The program also highlighted many other programs/projects that improved the City's knowledge of energy-saving considerations. The City was successful in developing and adopting an Energy Conservation Policy.
  - Annual Projected GHG Reductions: 86 MTCO$_2$, equal to the removal of 17 passenger vehicles from the road
  - Project Cost: $25,000
  - Payback Period: 25 months
- **Future Steps**—going out to the hydro turbine project when the major upgrade project is ready to bid.

SOMERTON MUNICIPAL WATER, ARIZONA (FEBRUARY 21, 2012)

- **Facility Profile**—Somerton Municipal Water is located in the City of Somerton, Arizona, within Yuma County. Yuma County is situated in the southwest corner of Arizona close to the California and Mexico borders. The water treatment plant was built in 1985. In 1998 the plant carried out a major facility upgrade by installing a new 1.2 million gal storage tank and a new 100 hp booster pump. The drinking water treatment plant is called the Somerton Municipal Water System and it sources water from 3–300-feet deep wells.
- **Baseline Data**—the System serves a population of 14,267 residents and treats 2 MGD. Approximately 907,000 kWh is used to run the System and

it costs an average of $85,000 per year for electricity. The plant generated 512.79 MTCO$_2$ of GHG emissions.
- **Energy Improvement Management Plan**—water treatment plant staff started the energy improvement process by first evaluating the energy efficiency of existing wells and pumps. Upon inspection, repairs were made to two wells and three pumps. By fixing the wells and pumps, Somerton Municipal Water Systems will save $27,000 each year on the electricity bill. Moreover, they will save an average 250,000 kWh annually which will result in a reduction in CO$_2$ emissions by an estimated 172 MTCO$_2$.
- **Challenges**—it was difficult for staff to find the time to participate in the Energy Management sessions. No additional staff resources were available to complete the project.
- **Accomplishments**—Somerton staff gained a better understanding of how projects are selected and increased their effectiveness in working with management. By upgrading wells and booster pumps they were able to save approximately $27,000 per year per pump on energy costs.
  - Annual Projected GHG Reductions: 172 MTCO$_2$, equal to the removal of 34 passenger vehicles from the road
  - Project Cost: $131,203, with $33,500 covered by incentives
  - Payback Period: 4 years
- **Future Steps**—project completion; begin solar project that will produce 1.5 million kWh.

HAWAII COUNTY DEPARTMENT OF WATER SUPPLY (FEBRUARY 21, 2012)

- **Facility Profile**—the Hawaii County Department of Water Supply (DWS) is one of the 21 departments within the County of Hawaii. The Country encompasses the entire island of Hawaii, and the administrative offices are located in Hilo, Hawaii. The DWS is the public potable water distribution utility; however, the island of Hawaii also has several other water systems that are not owned and operated by DWS. Hawaii, the largest island in the Hawaiian chain, is 93 miles long and 76 miles wide with a land area of approximately 4,030 square miles. The DWS operates and maintains 67 water sources and almost 2,000 miles of water distribution pipeline. Over 90 percent of the water served is from a groundwater source that requires minimal treatment with chlorine for disinfection.
- **Baseline Data**—the DWS serves 41,507 customers and produces 31.1 MGD. Because of the vast area, mountainous terrain of the island, and the many separate water sources, the energy needed to pump water to the surface is significant and plant employees drive a significant distance to operate the water distribution system. In 2010, the plant spent $16.5 million on energy costs, used 54,781,373 kWh for electricity, and 95,100 gal of gas and diesel. Hawaii County DWS emits an estimated 31,784.35 MTCO$_2$ of GHG emissions.
- **Energy Improvement Management Plan**—Hawaii County DWS energy reduction strategy's main goal is to reduce gas and diesel consumption.

Energy Efficient Operating Strategies

The performance target was to reduce fuel purchases by 200 gal per month or 2,400 gal annually. The target was met by establishing more efficient operators' routes around Hilo, using an automated SCADA system, and installing GPS equipment in 30 vehicles. A new automated SCADA system replaced the manual system. By reducing fuel consumption and increasing efficiency, DWS will reduce CO_2 emissions by an estimated 17.3 TCO_2 per year. This project was fully implemented on December 31, 2011. DWS also chose to develop an Energy Policy.

- **Challenges**—introducing a new automated SCADA system (to replace the old manual system) required programming time, and duplicate systems until the new system was proven. The data being collected was all manual until the new vehicle equipment was installed. DWS decided to purchase vehicle GPS units, so a new vehicle policy was needed. The pilot project modified city of Hilo operators' routes. The pilot project covers about one-third of the island. Implementing route changes met resistance because operators lost overtime. Establishing an Energy Management team was unsuccessful so the project was implemented by one person, but there were many moving parts.
- **Accomplishments**
  - Created an Energy Policy
  - Annual Projected Energy Savings: 1,965 gal of gasoline
  - Annual Overtime Savings: $9,780
  - Annual Projected Cost Savings: $17,670
  - Annual Projected GHG Reductions: 18 $MTCO_2$, equal to the removal of 3.4 passenger vehicles from the road
  - Project Cost: $25,300
  - Payback Period: 1.5 years
- **Future Steps**
  - Complete purchase of vehicle GPS systems.
  - Fully implement the project throughout the island, which will include 100 vehicles.
  - Begin a wind power project to generate renewable energy.

Eastern Municipal Water District, California (February 21, 2012)

- **Facility Profile**—Eastern Municipal Water District (EMWD) has over 250 operating facilities that have been constructed over the last 60 years. EMWD selected a drinking water filtration plant for the Energy Management Initiative. Perris Water Filtration Plant is located in Perris, California, within Riverside County. Perris is situated southeast of Los Angeles along the Escondido Freeway. The plant treats water pumped from the Colorado River and/or from the CA State Water Project.
- **Baseline Data**—Perris Water Filtration Plant:
  Water Treatment Design Capacity: 24 MGD
  Annual Energy Consumption: 5.6 million kWh
  Annual Cost of Energy: $695,000

GHG Emissions: 3,862 MTCO$_2$
Design Average Flow (FY 2011): 10.2 MGD
- **Energy Improvement Management Plan**—EMWD chose to reduce their energy consumption and greenhouse gas emissions by producing renewable energy on-site. The municipality will install a renewable power generator (in-conduit hydro generation) at the Perris Water Filtration Plant. The renewable power generator will produce up to 290,000 kWh of energy each year. The project will take one year to complete once funding is secure and will cost approximately $350,000.
- **Challenges**—these include identifying hydro-generation technology capable of meeting the low head pressure and varying flow conditions existing at the Perris Water Filtration Plant. These were technical challenges that were eventually overcome. Also, a grant application for funding from the U.S. Bureau of Reclamation was not funded, but EMWD obtained feedback on the proposal and will fund the project without a grant.
- **Accomplishments**
  - Raw water supply to EMWD's Perris Valley Water Filtration Plant is controlled through an existing valve which requires frequent, and costly, replacement. The benefits of this project are that it will eliminate the need for this replacement, and provide the necessary flow control capabilities combined with energy generation. EMWD has contracted and completed a feasibility study which shows the viability of the proposed project. The project which is capable of producing nearly 300,000 kWh of electricity is now past the design phase. Included in the design are revised cost estimates and inclusion of hydro turbine technology that meets the unique challenges of this application (low head, with highly variable flow).
  - EMWD has increased the awareness of systematic processes for analyzing overall energy management efforts, and provided structure and a strategic approach to energy management as a whole.
    - Annual Projected Energy Savings: 290,000 kWh
    - Annual Projected Cost Savings: $36,000
    - Annual Projected GHG Reductions: 200 MTCO$_2$, equal to the removal of 39 passenger vehicles from the road
    - Project Cost: $350,000
    - Payback Period: 5 years (factoring in the need to replace an existing, non-energy generating valve); 10 years (if the valve hadn't needed to be replaced)
- **Future Steps**—board approval of funding, go out to bid

Port Drive Water Treatment Plant, Lake Havasu, Arizona (February 21, 2012)

- **Facility Profile**—Port Drive Water Treatment plant (PDWTP) is located in Mohave County, Arizona, and serves the majority of the population of Lake

Energy Efficient Operating Strategies

Havasu City. The City is located along the Colorado River on the eastern shores of Lake Havasu. The water treatment plant was built between 2002 and 2004 and has never received a major facility upgrade. Currently, an estimated 86 percent of the water treated is sourced from groundwater wells and a small percentage comes directly from Lake Havasu.

- **Baseline Data**—PDWTP uses a natural biological process to remove iron and manganese from 11 million gal of water per day. The treated drinking water is then distributed to 50,000 customers. Annually, the plant uses 6,636,960 kWh of energy at a cost of $612,749 to the City each year. This equates to 4,577 $MTCO_2$ of GHG emissions.
- **Energy Improvement Management Plan**—Lake Havasu City Water Treatment Plant staff, encouraged by the EPA, completed a test "Change of Operations" of the North and South High Service Pump Station.
- **Challenges**—overall, since the 2009 recession, one major challenge to any non-core activity—including the effort—has been a lack of personnel resources. Despite these challenges, the City was able to implement small incremental changes. A major benefit for the City was the realization that those changes can and will be valuable in the future. This was seen in the demonstration project described in more detail in the accomplishments section. The lack of staff hours prevented the City from moving forward with an Energy Improvement Plan. This also delayed the implementation of a pump study until November 2011 which should result in some energy savings in 2012.
- **Accomplishments**
    - The City changed how much and when water would be released at each lead pump. The lead pump for each system was changed to operate at full water flow until it reached the turn-off-level setpoint. Prior to this, the variable frequency drive (VFD) controller was programmed to slow the lead pump at a nearly full-tank level to provide continuous flow through the WTP process. This change resulted in an average 8.7 percent energy reduction in pump stations for the months of March, April, and May.
    - For the months of June, July, and August, the average energy reduction was 6 percent for the north pumps, and 6.7 percent for the south, compared to the same months in 2010. There was no change in the water quality or ability to supply water on demand with these changes.
    - All light fixtures were replaced.
    - In the last 10–12 years, increase in electricity costs has not been passed on to users. Last year there was a 24 percent rate hike, but Lake Havasu reduced energy use by 30 percent.
    - In addition to the test project, Lake Havasu City qualified for an energy audit which was conducted for the North Regional Wastewater Treatment Plant by a consultant funded through the EPA. This resulting in a draft report submitted to Lake Havasu City in September 2011. The report identified seven future projects that may be scheduled in the

future. Many of the suggestions in their report may be relevant to Lake Havasu City's other treatment plants. Approximately 4 million gallons a day of wastewater is treated by these facilities.
- The anticipated energy savings of the Lake Havasu City Water treatment plant should equate to approximately 130,000 kWh annually.
  - Annual Projected GHG Reductions: 90 MTCO$_2$, equal to the removal of 18 passenger vehicles from the road
  - Project Cost: Zero
  - Payback Period: Zero
- **Future Steps**—Test savings of 6–8.7 percent during the next six months; implement energy audit recommendations. All parts of the plant are being examined for energy-saving opportunities.

TRUCKEE MEADOWS WATER AUTHORITY, RENO, NEVADA (FEBRUARY 21, 2012)

- **Facility Profile**—Truckee Meadows Water Authority (TMWA) has chosen its drinking water facility, Chalk Bluff Water Treatment Plant, for the Energy Management Initiative. The plant was built in 1994 and serves more than 330,000 customers throughout 110 square miles within Washoe County, Nevada. The Chalk Bluff Plant treats water from the Truckee River, which flows from Lake Tahoe and the Sierra Nevada Mountain range.
- **Baseline Data**—TMWA serves 93,000 customer connections. In 2009 the water authority spent just under $7 million on electricity and $88,000 on natural gas. The Chalk Bluff Water Treatment Plant uses 13.5 GWh (gigawatt hours) and 74,452 therms per year, and spends $1.3 million for electricity. The total energy use results in estimated annual GHG emissions of 9,309 MTCO$_2$.
- **Energy Improvement Management Plan**—while TMWA relies on gravity as much as possible, in a mountainous community, pumping water is a reality. The Chalk Bluff Plant is TMWA's largest water producer and the highest energy use facility. The high energy use is due to the pumping of water uphill from the river into the plant. To reduce energy consumption at Chalk Bluff, the implementation plant consists of two parts: (1) optimizing the time-of-use operating procedures and (2) water supply capital improvements.
  - The first strategy is to optimize time-of-use operating procedures by creating a mass flow/electric cost model of the treatment and effluent pumping processes. The model will be used to predict how changes to the operating procedure will affect electricity cost. In 2010, TMWA spent $938,000 on 7.8 GWh for non-water supply processes of the plant. This project intended to reduce non-supply electric costs by 15 percent or $141,000.
  - The second project involves water supply improvements to the Highland Canal which transports 90 percent of Chalk Bluff's water directly to the plant using gravity. The improvement plan will allow 100 percent of the water to be brought to the plant using the Highland Canal and meets

multiple objectives. Improvements will be made during winter months when customer water demand is the lowest to reduce the water supply pumping costs during construction. Currently, TMWA spends $60,000 on 0.5 GWh for water supply pumping at the Chalk Bluff Plant. Energy use will be zero when the project is complete. The design life of the new infrastructures is over 100 years and it will require no energy to operate.

- **Challenges**—originally scheduled to begin construction during the fall of 2011, delays in obtaining highway encroachment permits had postponed construction. To minimize water supply pumping costs during construction and therefore continue to reduce energy costs, this project has been delayed until the fall of 2012.

  TMWA attempted to use a mass balance/electric cost model to optimize time-of-use operating procedures. However, the mass balance/electrical cost model is not capable of matching the sophisticated decision-making capabilities of the experienced water plant operators. Therefore, the purpose of the model has shifted from generating decisions to being one of the several techniques used for improving time-of-use energy optimization at the Chalk Bluff Plant.

- **Accomplishments**
  - TMWA began setting and tracking time-of-use electricity goals for the Chalk Bluff Plant in November of 2010. The goals depend on time of day (e.g., 200 kW on-peak, 400 kW mid-peak, and 950 kW off-peak), and vary with season, based on the electric utility's tariffs. Water plant operators have the ability to be innovative in order to meet electricity use goals and system demands. The mass balance/electric cost modeling effort was valuable to establish baseline energy usage by (1) formally inventorying energy-intensive unit processes, (2) establishing kW draw of equipment, (3) establishing and ranking historic kWh usage of equipment, and (4) suggesting starting point kW targets for further optimization by operators.

    TMWA considers the time-of-use optimization project a great success due to its ability to save energy costs, and will continue to optimize and track the project's results. For the 12 months from November 2010 through October 2011 the time-of-use optimization has saved more than $225,000 (24.4 percent) compared to the same period the previous year. During this time electric energy usage was reduced by only 0.45 GWh (5.8 percent), indicating the savings was primarily due to improved time-of-use cost management.

    Going through the process of identifying energy needs of each process was eye opening. Talking to the operators and getting them to work toward the time-of-use goals was educational, engaging, and strengthened the team.
    - Annual Projected GHG Reductions: 310 MTCO$_2$, equal to the removal of 61 passenger vehicles form the road
    - Project Cost: Zero

- Payback Period: Zero
- Design is substantially complete for the water supply improvement project, and highway encroachment permits are expected in time for the project to proceed in the fall of 2012.
  - Annual Projected GHG Reductions: 345 MTCO$_2$, equal to the removal of 68 passenger vehicles from the road
  - Project Cost: $3,000,000
  - Payback Period: 50 years
- **Future Steps**—(1) complete project; continue to optimize and track project results and (2) get permits; construction.

TUCSON WATER, ARIZONA (FEBRUARY 21, 2012)

- **Facility Profile**—Tucson Water provides clean drinking water to a 330 square-mile service area in Tucson, Arizona. Located in Pima County, Tucson is positioned along Highway 10 about 70 miles north of the U.S./Mexico border. The potable water system serves roughly 85 percent of the Tucson metropolitan area, serving 228,000 customers. The potable system includes 212 production wells, 65 water storage facilities, and over 100 distribution pumps. A separate reclaimed water system serves parks, golf courses, and other turf irrigation. Tucson Water sources drinking water through groundwater and the Colorado River, by recharging and recovering river water delivered through the Central Arizona Project. Tucson Water uses recycled water for its reclaimed water system.
- **Baseline Data**—Tucson Water spends on average $12.5 million on the energy to operate its portable system and produces approximately 110 MGD of potable water per year. Annually, Tucson Water uses approximately 115,000,000 kWh and 5,000,000 therms of energy to run the system. This energy use results in an estimated 83,367.43 MTCO$_2$ of GHG emissions.
- **Energy Improvement Management Plan**—Tucson Water is partnering with Tucson Electric Power (TEP), a privately held, regulated electric utility, to reduce peak demand. TEP contracted EnerNOC (which develops and provides energy management applications and services for commercial, institution, and industrial customers, as well as electric power grid operators and utilities) to facilitate a new demand management program to reduce the energy load during peak hours. EnerNOC will pay customers to shed the load. Tucson Water will participate by identifying sites that are appropriate for load shedding. It has been estimated that the program will save Tucson Water 24,000 kWh during peak energy use and created $9,000 in offsetting revenue to be put toward energy cost.

As part of the City of Tucson's award under the Department of Energy's Energy efficiency and Conservation Block Grant (funded by the American Recovery and Reinvestment Act), Tucson Water is implementing a Water System Distribution Pump Efficiency Project. The project is designed to establish baseline data and data management tools for system booster pumps, provide energy-savings recommendations for the distribution

Energy Efficient Operating Strategies 93

system, and implement prioritized energy-savings upgrades. In addition, training will be provided and results from the project will provide actionable information on the cost effectiveness of continuing a program without grant funding. Projected energy and cost savings for the project are 350,000 kWh and $30,000 (year one, past project), respectively. The project is scheduled to be completed in early fall of 2012.
- **Challenges**—it was difficult to complete the project within one year. A two-year program would have given more time to implement the project in our Energy Management Plan.
- **Accomplishments**
  - In addition to the projected $9,000 cost savings of the EnerNOC program, energy data will inform any plans to expand real-time energy monitoring. At the end of the grant-funded booster pump project, the utility will realize energy and cost savings and have the information necessary to scope a more permanent pump efficiency program.
    – Annual Projected Energy Savings: 24,000 kWh during peak energy use periods
    – Annual Projected Cost Savings: $9,000
    – Annual Projected GHG Reductions: None
    – Project Cost: Zero
    – Payback Period: Zero
- **Future Steps**—the peak demand project is complete. Tucson will continue implementing an ARRA funded Pump Efficiency project that is estimated to save $30,000 and 350,000 kWh/year.

CHINO WATER PRODUCTION FACILITY, PRESCOTT, ARIZONA (FEBRUARY 21, 2012)

- **Facility Profile**—the Prescott-China Water Production Facility is located within Yavapai County in the town of Chino Valley, Arizona. Prescott is positioned close to the center of Arizona between Phoenix and Flagstaff just outside the Prescott National Forest. The Facility consists of a production well field, reservoir, and booster pump facility. The Facility was built in 1947 and received its last major facility upgrade in 2004.
- **Baseline Data**—the Facility supplies 50,000 residents with drinking water. During the winter the plant treats 4.5 MGD of water and peaks at 12 million gallons during the summer. The cost of electricity for the plant is $1,600,000 for 11,000,000 kWh/year. Annual GHG emissions are 7,581 $MTCO_2$.
- **Energy Improvement Management Plan**—the Facility has wells that are 15 miles north and lower in elevation than the treatment plant. The challenge was to reduce the $2,000/month demand charge. The Energy Management Plan for the Facility includes replacing the existing step voltage starts on three wells with soft-start units. The soft-start units will reduce the instantaneous demand on the power supply which will reduce the demand charge on the utility bill. It will also help to reduce the power surge on the power distribution system. In addition, softer starters help to extend the life of the well motors. Overall, the project saves money and

electricity through reduced demand charges, energy use, and maintenance. The estimated immediate cost saving associated with the reduced demand is $1,260/month, for a total savings of $15,120/year. The savings toward the maintenance and reduced strain on the electrical distribution system will be a long-term progressive savings.
- **Challenges**—the main challenge has been the loss of two members of the City's Energy Management team, which reduced management support and buy-in from staff. This delayed the ultimate implementation and construction of the soft-start project.
- **Accomplishments**—as the City neared the end of the program, the project gained acceptance. The City has completed a specification package and will soon advertise for construction. The energy management program has promoted an awareness of the City's energy use and potential savings associated with minor or major changes to operations. It has provided a good networking opportunity to learn, expand concepts, and consider new options. This project will reduce direct electrical costs, long-term maintenance costs, and result in unseen benefits like improved safety due to reduced instantaneous electrical demands. The Energy Management Programs also highlighted many other programs/projects that improved the City's awareness of energy savings. The City has implemented a new Energy Conservation Policy.
  - Annual Projected GHG Reductions: None
  - Project Cost: $42,000
  - Payback Period: 34 months
- **Future Steps**—advertise for construction, and begin a 2 MW solar generation project to offset 30 percent of water booster costs.

SOMERTON MUNICIPAL WASTEWATER TREATMENT PLANT, ARIZONA (FEBRUARY 21, 2012)

- **Facility Profile**—the Somerton Municipal Wastewater Treatment Plant is located in the city of Somerton within Yuma County, Arizona. Somerton is positioned along Highway 95, close to the border of California and Mexico. The wastewater treatment plant was built in 1985 and received its last major facility upgrade in 2011, when it was changed from a Sequencing Batch Reactor (SBR) facility with a capacity to treat 0.8 MGD to a MLE (Modified Ludzack-Ettinger) process with a capacity of 1.8 MGD.
- **Baseline Data**—Somerton Municipal Wastewater Treatment Plant serves a population of 14,296 residents. The wastewater treatment plant treats a daily average of 0.750 MGD of wastewater each day, and uses approximately 744,480 kWh/year, that generates 32 MTCO$_2$ of GHG emissions.
- **Energy Improvement Management Plan**—the objective of Somerton Municipal Wastewater Treatment plant is to save energy by running a more efficient plant. The City achieved this by replacing four old blowers used to run the 4-tank aeration system with two new more efficient blowers. (Current system only needs 1 new blower.) The City also replaced an old diffuser system with one high efficiency diffuser. In addition, two old

blowers were replaced, one for the digester and one spare, with one high efficiency blower. Overall, the upgrades are expected to save the wastewater treatment plant 10 percent on their annual electricity bill and reduce their electricity use by 6 percent, while doubling the capacity of the facility.
- **Challenges**—originally the Energy Improvement Management Plan called for doubling the number of tanks in the existing SBR system. It was later determined that by using more efficient diffusers and Turbo blowers, the same number of tanks could be kept by changing the process. This resulted in keeping the same footprint and more than doubling capacity while reducing energy use by 10 percent.
- **Accomplishments**—plant energy use awareness increased. Even though the expansion is still not complete, the plant succeeded in reducing electricity usage by 12 percent and costs by 11.4 percent, while doubling treatment capacity.
  - Annual Projected Cost Savings: $29,328
  - Annual Projected GHG Reductions: 51 MTCO$_2$, equal to the removal of 10 passenger vehicles from the road
  - Project Cost: $146,640
  - Payback Period: 1 year
- **Future Steps**—begin a 1.5 million kWh solar project

REFERENCES AND RECOMMENDED READING

California Energy Commission. 2012. Managing Your Electrical Load. In: EPA, *Water & Energy Efficiency in Water and Wastewater Facilities.* Accessed 04/21/2020 @ http://epa.gov/reion09/waterinfrastructure/technology.html.

City of Portland, Oregon. 1994. *CSO Facilities Plan.* Prepared by CH2M Hill.

EPA. 1999. *Combined Sewer Overflow Technology Fact Sheet: Inflow Reduction.* 832-F-99-035. Washington, DC: Environmental Protection Agency.

EPA. 2012. *Water & Energy Efficiency in Water and Wastewater Facilities.* Accessed 04/28/2020 @ http://epa.gov/region09/waterinfrastrucutre/technology.html.

Hides, S. 1997. *Quantity is the Key to Improving Quality: A Common Sense Approach for Reducing Wet Weather Impacts.* Presented at the 26th Annual WEAO Technical Symposium and OPCEA Exhibitions, London, ON, Canada, July 1997.

Improving the Energy Efficiency of Wastewater Treatment Facilities. 1993. WSEO-192. Washington State Energy Office, June.

McKelvie, S. 1996. Flowslipping: An Effective Management Technique for Urban Runoff. *Proceedings from Urban Wet Weather Pollution Conference*, Quebec City, QC. Parson Brinckerhoff Quade & Douglas, Inc.

Spellman, F.R. 2007. *Handbook of Water and Wastewater Treatment Plant Operations.* 2nd ed. Boca Raton, FL: CRC Press.

7 Energy Conservation Measures for Wastewater Treatment

Providing reliable wastewater services and safe drinking water is a highly energy-intensive activity in the United States. A report prepared for the Electric Power Research Institute (EPRI) in 1996 estimated that by the end of that year, the energy demand for the water and wastewater industry would be approximately 75 billion kilowatt hours (kWh) per year, or about 3 percent of the electricity consumed in the U.S.

Franklin L. Burton, 1996

INTRODUCTION

In this chapter it is only fitting that a description of energy conservation measures (ECMs) that have been discussed briefly in the preceding chapters is expanded and discussed with regard to specifics for actual application in the wastewater treatment industry. I chose to focus on the wastewater industry in particular because it is a larger user of energy, much more so than the water treatment industry. However, the reader should understand that the energy conservation measures discussed in the following can also apply to water treatment operations. The purpose of this chapter is to encourage the implantation of ECMs at publicly owned treatment works (POTWs) by providing accurate performance and cost/benefit information for such projects.

Along with being familiar with public utility (water or wastewater) nomenclature, it is important for public utility managers to understand that energy is used throughout the wastewater treatment process; however, pumping and aeration operations are typically the largest energy users. Energy costs in the wastewater industry are rising due to many factors, including (USEPA 2008):

- Implementation of more stringent effluent requirements, including enhanced removal of nutrients and other emerging contaminants of concern that may, in some cases, lead to the use of more energy-intensive technologies
- Enhanced treatment of biosolids including drying/pelletizing
- Aging of wastewater collection systems that results in additional inflow and infiltration, leaning to higher pumping and treatment costs
- Increase in electricity rates

As a consequence of these rising costs, many wastewater facilities have developed energy management strategies and implemented energy conservation measures.

Accordingly, the chapter discusses and describes ECMs being employed in wastewater treatment plants by:

- Providing an overview of conventional ECMs related to pumping design, variable frequency drives (VFDs), and motors
- Describing innovations in ECMs related to blower and diffuser equipment. It includes a summary of various blower types such as single-stage centrifugal, high-speed turbo, and screw compressors in addition to new diffuser technology
- Providing a discussion of ECMs for advanced technologies (UV disinfection, membranes, and anoxic zone mixing) and present full-scale plant test results where available (USEPA, 2010)

PUMPING ENERGY CONSERVATION MEASURES

As mentioned earlier, pumping operations can be a significant energy draw at wastewater treatment plants (WWTPs), in many cases second only to aeration. Pumps are used for many applications. At the plant headworks, they may be used to provide hydraulic head for the downstream treatment processes. Within the plant, they are used to recycle and convey waste flows, solids, and treated effluent to and from a variety of treatment processes. Pumps are also found in remote locations in the collection system to help deliver wastewater to the plant.

The overall efficiency of a pumping system, also called the "wire-to-water" efficiency, is the product of the efficiency of the pump itself, the motor, and the drive system or method of flow control employed. Pumps lose efficiency from turbulence, friction, and recirculation with the pump (WEF 2009; Spellman & Drinan, 2001). Loss is also incurred if the actual operating condition does not match the pump's best efficiency point (BEP). The various mechanisms for controlling flow rate decrease system efficiency. Throttling valves used to reduce the flow rate increases the pumping head, flow control values burn head produced by the pump, recirculation expends power with no useful work, and VFDs produce a minor amount of heat. Of these methods, VFDs are the most flexible and efficient means to control flow despite the minor heat loss incurred. Table 7.1 summarizes typical pump system efficiency values—note that inefficiency in more than one component can add up quickly, resulting in a very inefficient pumping system.

DID YOU KNOW?

BEP is the flow rate (typically in gallons per minute or cubic meters per day) and head (in feet or meters) that gives the maximum efficiency on a pump curve.

Inefficiencies in pumping often come from a mismatch between the pump and the system it serves due to improper pump selection, changes in operating conditions, or

TABLE 7.1
Pump System Efficiency

Pump System Component	Range	Efficiency (%) Low	Average	High
Pump	30–85	30	60	75[1]
Flow Control[2]	20–98	20	60	98
Motor[3]	85–95	85	90	95
Efficiency of System		5.5	32	80

[1] For pumping wastewater. Pump system efficiencies for clean water can be higher.
[2] Represents throttling, pump control valves, recirculation, and VFDs.
[3] Represents nameplate efficiency and varies by horsepower.

the expectation that the pump will operate over a wide range of conditions. Signs of an inefficient pumping system include:

- Highly or frequently throttled control values
- Bypass line (recirculation) flow control
- Frequent on/off cycling
- Cavitation noise at the pump or elsewhere in the system
- A hot running motor
- A pump system with no means of measuring flow, pressure, or power consumption
- Inability to produce maximum design flow

The literature provides several examples of plants reducing pumping energy by as much as 5 percent through pump system improvements (Focus on Energy 2006). Energy savings result from lowering of pumping capacity to better match system demands, replacing inefficient pumps, selecting more efficient motors, and installing variable speed controllers. Generally speaking, energy conservation measures for pumping are conventional and do not represent an area where recent technological innovation has played a part in improving energy conservation and efficiency. Pumping ECMs are, however, still extremely important to reducing and optimizing energy use at wastewater treatment plants.

PUMPING SYSTEM DESIGN

Appropriate sizing of pumps is key to efficient operation of wastewater treatment plants. Pumps sized for peak flow conditions that occur infrequently or, worse, in the future toward the end of the pump's service life operate the majority of the time at a reduced flow that is below their BEP. Peak flow is typically several times greater than the average daily flow and can be an order of magnitude different than minimum flow; especially for small systems or systems with significant inflow and infiltration (I&I) in some systems, these projected future flows are never reached during the design life of the pump.

For existing treatment plants, utilities should evaluate the operation of existing pumps and identify opportunities for energy reduction. A good starting point is to determine the efficiency of existing pumping systems, focusing first on pumps that operate for the most hours and have potential problems identified earlier as pumping system inefficiencies. Plants should collect performance information on the flow rate, pressure, and delivered power to the pumps. Field measurements may be necessary if the plant does not regularly record this information. Pump and system curves can then be constructed to determine the actual operating points of the existing system. Operating points more than 10 percent different than the BEP signal room for improvement.

To improve efficiency, utilities should consider replacing or augmenting large capacity pumps that operate intermittently with smaller capacity pumps that will operate for longer periods and closer to their BEP. When replacing a pump with a smaller unit, both the horsepower and efficiency change. A quick way to estimate the annual energy cost savings is to approximate cost before and after the improvement and detriment the difference using equation (7.1):

$$\text{Annual Energy Savings } (\$) = \left[hp_1 \times L_1 \times 0.746 \times hr \times E_1 \times C \right] \\ - \left[hp_2 \times L_2 \times 0.746 \times hr \times E_2 \times C \right] \quad (7.1)$$

Where:
 hp_1 = horsepower output for the larger capacity pump
 hp_2 = horsepower output for the smaller capacity pump
 L_1 = load factor of the larger capacity pump (percentage of full load/100—determined from pump curve)
 L_2 = load factor of smaller capacity pump (percentage of full load/100—determined from pump curve)
 hr = annual operating hours
 C = energy (electric power) rate ($/kWh)
 E_1 = efficiency of the larger capacity pump
 E_2 = efficiency of the smaller capacity pump

See Example 7.1 for how the Town of Trumbull was able to save more than $1,500 per year by adding a small pump to one of its existing sewage pumping stations. When applied correctly, replacement of standard drives with VFDs can also yield significant improvements.

Example 7.1 Town of Trumbull, CT, Improves Efficiency at Reservoir Avenue Pump Station

Background: Wastewater from the Town of Trumbull, in southwestern CT, is collected and conveyed to a WWTP in Bridgeport via 10 sewage pump stations. One of these, the Reservoir Avenue Pump station, consisting of two 40 hp direct-drive pumps designed to handle an average daily flow of 236 gpm (gallons per minute). Each pump was operated at a reduced speed of 1320 rpm at 50.3 feet of total dynamic head (TDH) with a duty point of approximately 850 gpm. A bubbler-type level control system was used to turn the pumps off and on. One pump can handle the entire peak inflow (usually < 800 gpm) with the second pump operating only during peak flow conditions.

Energy Efficiency Upgrades: To reduce energy use, the town installed a new 10-hp pump and modified the system control scheme. The new pump handles the same volume as the original pump but operates for a longer time between standby periods. In addition, the speed control was eliminated and the original pumps, when used, are run at full speed of 1750 rpm. This allowed the impellers of the original pumps to be trimmed from 11.25 inches in diameter to 10 inches. The original pumps are used for infrequent peak flows that cannot be handled by the new 10 hp pump. Under normal operating conditions, the operating point for the new pump is 450gpm at 407 TDH compared to 850 gpm at 50.3 feet of head for the whole system. Improvements were made to the lighting and control systems, resulting in additional energy savings.

Energy Savings: Annual energy savings were 17,643 kWh from modifying the pumping system. Total energy savings were 31,875 kWh/year, or approximately $2600/year based on a rate of 8¢/kWh. Total implementation costs were $12,000, resulting in a simple payback of 4.6 years (USDOE, 2005b).

For Greenfield plants and/or new pump stations, utilities should consider and plan for staging upgrades of treatment capacity as part of the design process. For example, multiple pumps can be specified to meet a future design flow instead of one large pump so that individual pumps can be installed as needed, say at year zero, year ten, and year twenty. The State of Wisconsin's Focus on Energy best practices guidebook (Focus on Energy 2006) estimates that staging of treatment capacity can result in energy savings between 10 percent and 30 percent of total energy consumed by a unit process.

Pump Motors

The induction motor is the most commonly used type of alternating current (A-C) motor because of its simple, rugged construction and good operating characteristics. It consists of two parts: the **stator** (stationary part) and the **rotor** (rotating part). The most important type of polyphase induction motor is the three-phase motor.

Important Note: A three-phase (3-θ) system is a combination of three single-phase (1-θ) systems. In a 3-θ balanced system, the power comes from an alternating current generator that produces three separate but equal voltages, each of which is out of phase with the other voltages by 120°. Although 1-θ circuits are widely used in electrical systems, most generation and distribution of A-C current use 3-θ systems.

The driving torque of both direct current (D-C) and A-C motors is derived from the reaction of current-carrying conductors in a magnetic field. In the D-C motor, the magnetic field is stationary and the armature, with its current-carrying conductors, rotates. The current is supplied to the armature through a commutator and brushes. In induction motors, the rotor currents are supplied by electromagnet induction. The stator windings, connected to the A-C supply, contain two or more out-of-time-phase currents, which produce corresponding magnetomotive forces (mmfs). These mmfs establish a rotating magnetic field across the air gap. This magnetic field rotates continuously at constant speed regardless of the load on the motor. The stator winding corresponds to the armature winding of a D-C motor or to the primary winding of a transformer. The rotor is not connected electrically to the power supply.

The induction motor derives its name from the fact that mutual induction (or transformer action) takes place between the stator and the rotor under operating conditions. The magnetic revolving field produced by the stator cuts across the rotor conductors, inducing a voltage in the conductors. This induced voltage causes rotor current to flow. Hence, motor torque is developed by the interaction of the rotor current and the magnetic revolving field.

The cost of running these electric induction motors (and other motors) can be the largest fraction of a plant's total operating costs. The Water Environment Federation (WEF) estimates that electrical motors make up 90 percent of the electric energy consumption of a typical wastewater treatment plant (WEF 2009). Inefficient motors, operation outside of optimal loading conditions, and mechanical or electrical problems with the motor itself can lead to wasted energy at the plant and are opportunities for savings.

The percent energy savings resulting from replacing older motors with premium motors is modest, typically between 4 percent and 8 percent (NEMA Standard MG-1. 2006). Savings can be higher when energy audits reveal that exiting motors achieve very low efficiencies, or when existing motors are oversized and/or underloaded. Many plants have coupled motor replacements with upgrades from fixed speed to variable speed drives for significantly higher energy savings.

What is the bottom line on pump motors in wastewater treatment plant applications? Keep in mind that upgrading of motors is a conventional ECM that has been practiced in wastewater treatment plants for some time.

Motor Efficiency and Efficiency Standards

Motor efficiency is a measure of mechanical power output compared to electrical power input, expressed as a percentage.

$$\text{Motor efficiency} = P_m / P_e \tag{7.2}$$

Where:
P_m = mechanical power output of the motor in Watts
P_e = electrical power input to the motor in Watts (WEF 2009)

No motor is 100 percent efficient—all motors experience some power loss due to friction, electrical resistance losses, magnetic core losses, and stray load losses. Smaller motors generally experience higher losses compared to larger motors.

The U.S. Congress, in the Energy Policy Act (EPACT) of 1992, set minimum efficiency standards for various types of electric motors manufactured in or imported to the United States. Minimum nominal, full-load efficiencies typically range from 80 percent to 95 percent depending on size (i.e., horsepower) and other characteristics. Motors manufactured since 1997 were required to comply with EPACT standards and to be labeled with a certified efficiency value.

The National Electrical Manufacturers Association (NEMA) premium efficiency standard has existed since 2001 (NEMA 2006) as a voluntary industry standard and has been widely adopted due to its power (and thus cost) savings over EPACT 1992 compliance standards. The 2007 Energy Act raised efficiency standards of motors to NEMA Premium efficiency levels and set new standards for motors not covered by previous legislation.

Submersible motors are commonly used in wastewater treatment plants. They serve specialized applications in environment that are not suited for NEMA motors. There is currently no efficiency stand for submersible motors and their efficiency is less than NEMA motors. Additionally, their power factor is usually lower. Their selection is usually driven by the application, though some applications have alternatives that use NEMA motors. Efficiency should be considered in the evaluation of alternatives in these applications as it affects the life-cycle cost used in the selection process.

Operating efficiency in the field is usually less than the nominal, full-load efficiency identified by the motor manufacturer. One reason for this is the operating load. As a rule of thumb, most motors are designed to operate between 50 percent and 100 percent of their rated load, with maximum efficiency occurring at about 75 percent of maximum load. For example, a motor rated for 20 hp should be operated between 10 hp and 20 hp and would have its best efficiency around 15 hp. Larger motors can operate with reasonable efficiency at loads down to the 25 percent range (USDOE 1996). Motors operated outside of the optimal loading lose efficiency. Other factors that reduce efficiency in the field include power quality (i.e., proper voltage, amps, ad frequency) and temperature. Motors that have been rewound are typically less efficient compared to the original motor.

Accurately determining the efficiency of motors in service at a plant is challenging because there is no reliable field instrument for measuring mechanical output power. Several methods are available, however, to approximate motor efficiency.

Motor Management Programs

Wastewater utilities should consider purchasing new energy efficient premium motors instead of rewinding older units when replacing equipment and when making major improvements at the plant. Motor replacement is best done as part of a plant-wide motor management program. A first step in program development is to create an inventory of all motors at the plant. The inventory should contain as much information as possible, including manufacturers' specifications, nameplate information, and field measurements such as voltage, amperage, power factor, and operating speed under typical operating conditions. Following the data gathering phase, plant managers should conduct a motor replacement analysis to determine which motors to replace now and which are reasonably efficient and can be replaced in the future or at the time of failure.

> **DID YOU KNOW?**
>
> **Plants Should Consider Buying New Energy Efficient Motors:**
>
> - For new installations
> - When purchasing new equipment packages
> - When making major modifications to the plant
> - Instead of rewinding older, standard-efficiency units
> - To replace oversized and/or underloaded motors
> - As part of a preventive maintenance or energy conservation program
>
> *Source: Motor Challenge Fact Sheet: Buying an Energy Efficiency Electric Motor.* Available at http://eere.energy.gov/industry/bestpracites/pdfs/mc-03 82.pdf.

A key input to any motor replacement analysis is economics. A simple approach is to calculate the annual energy savings of the new motor compared to the old unit and determine the payback period in years (in other words, when will the cumulative energy savings exceed the initial costs). The following simple equation can be used to determine annual energy savings:

$$\text{Annual Energy savings } (\$) = \text{hp} \times L \times 0.746 \times \text{hr} \times C \times (E_p - E_e) \qquad (7.3)$$

Where:
   hp = horsepower output of motor
   L = load factor (percentage of full load/100)
   0.746 = conversion from horsepower to kW units
   hr = annual operating hours
   C = energy (electric power) rate ($/kWh)
   E_e = existing motor efficiency as a percentage
   E_p = premium motor efficiency as a percentage

Simple payback in years can then be calculated as the new motor cost (capital plus installation) divided by the annual energy savings. When comparing buying a premium motor instead of rewinding an existing one, the cost of rewinding the exiting motor should be subtracted from the motor cost. Any cash rebated from your local electric utility or state energy agency should also be subtracted from the cost of the new motor. When replacing pumps, motors, or control systems, upgrading the electrical service, wiring, transformers, and other components of the electrical system should be considered in calculating energy savings and life-cycle costs. Utilities should also consider the importance of reliability and environmental factors when making motor replacement decisions. More robust economic analysis such as net present value life-cycle cost analysis should be considered, especially for large expenditures.

The ENERGY STAR® Cash Flow Opportunity (CFO) calculator is an easy-to-use spreadsheet tool that can help plant managers calculate simple payback as well as cost of delay, which is the last opportunity cost if the project is delayed 12 months or more. The last sheet of the workbook provides a summary that can be given to senior managers and decision makers to help convince them of the financial soundness of energy efficiency upgrades. The CFO calculator and other financial tools are available for free download at http://energystar.gov/index.cfm?c=assess_value.financial_tools.

The task of motor inventory management and replacement analysis is made significantly easier by publically available software tools. Developed by the DOE Industrial Technologies Program, MotorMaster+ is a motor selection and management tool, available free online at http://motorsmatter.org/. It includes inventory management features, maintenance logging, efficiency analysis, savings evaluation, and energy accounting. It includes a catalog of 17,000 motors from 14 manufacturers, including NEMA Premium* software, a spreadsheet tool the sponsors of the Motor Decisions Matter campaign developed to assist plant managers with motor replacement/repair decision making. The tool is titled the "1*2*3 Approach to Motor Management" and is available for free download at http://motorsmatter.org/tools/123approach.html.

Innovative and Emerging Technologies

Siemens Energy and Automation in cooperation with the Copper Development Association has developed "ultra-efficient" copper rotor squirrel-cage-type induction A-C motors. These motors exceed NEMA Premium full-load efficiency standards by up to 1.4 percent; however, they are only currently available in outputs up to 20 hp. In addition to using high-conductivity copper rotors in place of aluminum, the new motors have the following efficiency improvements:

- Optimized rotor and stator design
- Low-friction bearings
- Improved cooling system
- Polyurea-based grease
- Dynamically balanced rotors
- Precision-machined mating surfaces for reduced vibration
- Motor insulation compatible with VFD (USDOE 2008)

The U.S. Department of Energy (USDOE), in cooperation with Baldor Electric Company and other private partners, is developing a new grade of Ultra-Efficient and Power-Dense Electric Motors, with the goal of a 15 percent reduction in motor energy loss over NEMA Premium motors. For example, if a NEMA Premium motor with particular characteristics and output horsepower was 92 percent efficient and thus had 8 percent loss, this new grade of motor would reduce loss by 0.15× 8 percent = 1.2 percent, for a new overall efficiency of 93.2 percent. The new grade of motor will also be 30 percent smaller in volume and 30 percent lower in weight, leading to decreased motor cost due to lower material costs (USDOE 2009).

Power Factor

Power factor is important because customers whose loads have low power factor require greater generation capacity than what is actually metered. This imposes a cost on the electric utility that is not otherwise recorded by the energy and demand charges. There are two types of power that make up the total or apparent power supplied by the electric utility. The first is *active* (also called *true* power) power. Measured in kW, it is the power used by the equipment to produce work. The second is the *reactive* power. This the power used to create the magnetic field necessary for induction devices to operate. It is measured in kVARs.

- Active Power (True Power) (P): power that performs work measured in Watts (W)
- Reactive Power: power that does not perform work (sometimes called "wattless power") measured in VA reactive (VAr)
- Complex Power: the vector sum of the true and reactive power measured in volt amps (VA)
- Apparent Power: the magnitude of the complex power measured in volt amps

The vector sum of the active and reactive power is called complex power. Power factor is the ratio of the *active* power to the *apparent* power. The power factor of fully loaded induction motors ranges from 80 percent to 90 percent depending on the type of motor and the motor's speed. Power factor deteriorates as the load on the motor decreases. Other electrical devices such as space heaters and old fluorescent or high discharge lamps also have poor power factor. Treatment plants have several motors, numerous lamps, and often electric heaters, which, combined, lowers the facility's overall power factor.

Power factor may be leading or lagging. Voltage and current waveforms are in phase in a resistive A-C circuit. However, reactive loads, such as induction motors, store energy in their magnetic fields. When this energy gets released back to the circuit it pushes the current and voltage waveforms out of phase.

Improving power factor is beneficial as it improves voltage, decreases system losses, frees capacity to the system, and decreases power costs where fees for poor power factor are billed. Power factor can be improved by reducing the reactive power component of the circuit. Adding capacitors to an induction motor is perhaps the most cost-effective means to correct power factor as they provide reactive power. Synchronous motors are an alternative to capacitors for power factor correction.

Variable Frequency Drives

VFDs are used to vary the speed of a pump to match the flow conditions. The speed of a motor is controlled by varying the frequency of the power delivered to the motor. The result is a close match of the electrical power input to the pump with the hydraulic power needed to pump the water. In this section the VFD discussion is expanded and includes energy savings, applications, and strategies for wastewater pumping stations.

Energy Savings

VFDs have been used by many wastewater utilities to conserve energy and reduce costs. A literature review found numerous success stories with energy savings ranging from 70,000 kWh/year for smaller WWTPs (i.e., average daily flow of 7–10 mgd) to 2,8000,000 kWh/year for larger WWTPs (i.e., average daily flow of 80 mgd) (EPRI 1998; Efficiency Partnership 2009; USDOE 2005c). VFDs are now more easily available and affordable, and paybacks for VFDs range from six months to five years depending on the existing level of control and annual hours of operations (Focus on Energy, 2006).

To approximate the potential energy savings, utilities should develop a curve of actual flow in hourly increments during a day. Using the curve, energy consumed by a constant-speed motor and throttling valve can be estimated and compared to energy consumed by a VFD system that matches the hourly flow rate to power used.

Applications

VFDs can be installed at remote collection system pumping stations, at lift stations, on blowers, and on oxidation ditch aeration rotor drives. A common application of VFDs is for pumps that experience a large variation in diurnal flow, such as at wastewater pumping stations. However, if VFDs are not selected and applied correctly, they can waste energy. Operating below 75 percent for full load, VFDs can have very low efficiencies. In selecting a VFD, information should be obtained from the VFD manufacturer showing the efficiency at different turn down rates.

VFDs are not applicable in all situations. VFDs may not be effective when a large static head must be overcome or where there is little variation in the flow rate (WEF 2009). Additionally, some motors are not suited for used with VFDs. When the drive reduces the frequency to the motor the voltage decreases. However, the amperage increases which can generate heat. More commonly, voltage spikes that develop from the non-sinusoidal wave form produced by VFDs can damage motor insulation if not properly filtered. Conductors within the motor should be properly insulated and the motors should be capable of dissipating the heat.

VFD Strategies for Wastewater Pumping Stations

VFDs can be costly to install in an existing pump station and require space in the electrical room. The range of flow, number of pumps, and hours of operation also need to be considered when evaluating the implementation of VFD control. Although equipping all pumps with VFDs provides maximum operation flexibility, this can be costly and, in retrofit projects, not always feasible. Often the rewards of having VFDs can be achieved at less cost with half or as few as one pump being equipped.

One VFD can be feasible in small stations where two pumps are run in duty/standby mode because the duty pump runs the majority of the time, reaping the savings with the VFD. In situations where both pumps are run in the lead/lag mode to cover the range of flow encountered it is usually beneficial to have both pumps equipped with VFDs. This allows the pumps to alternate the lead position, which balances their hours, and simplifies the controls as both pumps can be operated in the same manner.

In the case of larger stations with three or more pumps of the same size operated in lead/lag mode, the number of VFDs needed depends on the range of flow and the space available. If one pump runs the majority of the time with infrequent assistance from the others, then one VFD would likely suffice. However, if the second pump operates frequently, then at least two VFDs are recommended. In the two-VFD scenario, when an infrequent peak flow is needed, the third constant-speed pump can provide the base load while both VFD-driven pumps adjust to meet the demand. Depending on the size of the pumps, it could be more beneficial to install a smaller pump instead and run it with a VFD. This maximizes the efficiency of the system because when the large pumps are run, they are near there BEP without the heat losses generated by VFDs.

Large stations with multiple pumps of different sizes need to be evaluated on a case-by-case basis. Typically, VFDs are placed on the smaller pumps so that they can be used to fill in the peaks before another large pump is turned on. The controls are simple and sequencing is easy to maintain when a pump is down for service. Additionally, the cost is lower as small VFDs are less expensive than large ones.

It is important to run each pump periodically. Bearings in pumps that sit for too long can be damaged from brinnelling (i.e., a material surface failure caused by contact stress that exceeds the material limit; it occurs when just one application of a load is great enough to exceed the material limit; the result is a permanent dent or "brinnell" mark) and stuffing boxes can dry out and leak. It is beneficial from an operation and maintenance (O&M) standpoint to exercise equipment at intervals recommended by the equipment manufacturer to ensure their reliability when called upon. Energywise, it is best to do this during off-peak electrical hours such as morning or on weekends.

AERATION BLOWERS

In wastewater treatment the aeration process can account for 25 percent to as much as 60 percent of total plant energy use (WEF 2009); this is a reality that public utility managers need to know and usually find out to be a fact quickly during their tenure. This section provides technical information and cost/energy data for ECMs related to innovation and emerging blowers and diffuser equipment. Note that unlike other ECMs described in this book, blower and diffuser designs are often unique to manufacturers. Hence, this section contains information on proprietary systems as examples. The wastewater industry is constantly evolving and new equipment not identified in this book may be available or emerging in the future. When evaluating new equipment, design engineers and plant owners should work closely with their state regulatory agency to assess operating principles and potential energy savings

Blowers are an integral piece of the aeration system. There are many configurations, but all consist of lobes, impellers, or screws mounted on one or more rotating shafts powered by a motor. As the shaft turns, the blower pulls in outside air and forces it through distribution pipes into aeration basins at pressures typically 5–14 psig (pounds per square inch gauge). The energy consumption of blowers is a function of air flow rate, discharge pressure, and equipment efficiency (WEF 2009). Blower efficiency varies with flow rate, speed, pressure, inlet conditions, and actual design.

Blowers can be categorized as either (1) positive displacement blowers, which provide a constant volume of air at a wide range of discharge pressures, or (2) centrifugal blowers, which provide a wide range of flow rates over a narrow range of discharge pressure. Centrifugal blowers are either multi-stage, with a sequence of impellers mounted along a single shaft directly connected to a motor with a flexible coupling, or single-stage, with one impeller typically with speed increasing gears or a variable frequency drive. Single-state centrifugal blowers can be conventional integrally geared blowers or gearless (also known as high-speed "turbo") blowers. Positive displacement or centrifugal blowers (multi-stage or new high-speed turbo blowers) are well suited for small plants. Large plants more often use multi- or single-stage centrifugal blowers as high-speed turbo blowers are not yet available in capacities suitable for large plants. Table 7.2 lists the types of blowers, description of operation, and provides information on their advantages and disadvantages.

Table 7.3 presents typical ranges of isentropic (nominal) energy efficiency and turndown (low flow) for different blower types. Note that there is significant variation from small to large blowers of any type; the values presented are general rules of thumb and may vary with the application.

Controlling positive displacement blowers is typically done by varying blower speed with a variable frequency drive or with the use of multiple lowers operating in parallel. Throttling air flow through the machine is not possible for this type of blower. Multi-stage centrifugal blowers can be controlled through a variety of techniques, the most efficient being VFDs followed by suction air flow throttling using inlet butterfly valves. WER (2009) reports that VFD operation of multi-stage centrifugal blowers is 15–20 percent more efficient than throttling.

This section identifies several innovative and emerging ECMs related to blower and diffuser equipment:

- Turbo blowers are a significant area of innovation in blower design offering energy savings for the wastewater industry. They emerged in the North American market around 2007 and have been or are being tested and installed at many plants.
- Single-stage centrifugal integrally geared blowers are controlled using inlet guide vanes and variable diffuser vanes. This control technique has the advantages of managing air flow and pressure independently.
- Where fine bubble diffusers were once considered the standard for energy efficiency, new materials and configurations capable of producing "ultra-fine" bubbles (1 mm or less) are now available.
- Technological advances are also progressing in the area of diffuser cleaning.

A very new technology is the rotary screw compressor. The technology was released to the U.S. market in the summer of 2010. The manufacturers claim significant energy savings of up to 50 percent compared to rotary lobe blower technology. Units are being manufactured by Atlas Copco, AERZEN, Inc., and Dresser Roots.

TABLE 7.2
Overview of Blower Types for Aeration of Wastewater

Category	Description and Operation	Advantages	Disadvantages
Positive Displacement	Provides fixed volume of air for every shaft revolution. Operates over a wide range of discharge pressures	• Low capital cost, economical at small scale • Can achieve higher output pressure at same air flow rates • Simple control scheme for constant flow applications	• Difficult to operate at variable flow rates without VFD • Can be noisy (enclosures are commonly used for noise control) • Requires more maintenance than other types • Typically least energy efficient
Centrifugal Multi-Stage	Uses a series of impellers with vanes mounted on rotating shaft (typically 3,600 rpm). Each successive impeller increases discharge pressure. Individual units operate at narrow range of discharge pressures at wide range of flow rates	• Can be more energy efficient than positive displacement • Lower capital cost compared to single-stage centrifugal blowers • Can be quieter than single-stage units	• Can be less energy efficient than single-stage centrifugal • Efficient than single-stage centrifugal • Efficiency decreases with turndown
Centrifugal Single-Stage Integrally Geared	Similar to multi-stage but uses a single impeller operating at high speed (typically 10,000–14,000 rpm) to provide discharge pressure. Uses gearing between motor and blower shaft	• Can be more energy efficient than multi-stage or positive displacement • Can maintain good efficiency at turndown • Typically come with control system to surge protection	• More moving parts than multi-stage units. Surge can be more damaging • Can be noisy (enclosures are commonly used for noise control) • Higher capital cost compared to multi-stage or positive displacement
Centrifugal Single-Stage Gearless (High-Speed Turbo)	Centrifugal single-stage blower uses special low-friction bearings to support shaft (typically 40,000 rpm). Uses a single or dual impeller	• Small footprint • Efficient technology for lower air flow capacity • Can maintain good efficiency at turndown • May come with integrated control systems to modulate flow and for surge protection • Can be easy to install (place, plumb, and plug in)	• Typically higher capital cost compared to multi-stage or positive displacement blower (although likely less expensive than integrally geared) • Limited experience (new technology) • More units required for larger plants (will change as manufacturers expand air flow range)

Source: Adapted from WEF 2009, WEF and ASCE 2010.

TABLE 7.3
Typical Blower Efficiencies

Blower Type	Nominal Blower Efficiency (percent)	Nominal Turndown (percent of rated flow)
Positive Displacement (variable speed)	45–65	50
Multi-Stage Centrifugal (inlet throttled)	50–70	60
Multi-Stage Centrifugal (variable speed)	60–70	50
Single-Stage centrifugal, Integrally Geared (with inlet guide vanes and variable diffuser vanes)	70–80	65
Single-Stage Centrifugal, Gearless (High-Speed Turbo)	70–80	50

Source: Adapted from Gass, J.V. 2009. In Evaluation of Energy Conservation Measures. USEPA (2010).

HIGH-SPEED GEARLESS (TURBO) BLOWERS

High-speed gearless, or "turbo," blowers use advanced bearing design to operate at higher speeds (upward of 40,000 rpm) with less energy input compared to multi-stage and positive displacement blowers. Some turbo blowers come in package system with integrated VFDs and automated control systems to optimize energy efficiency at turndown.

Turbo blowers are available in two primary configurations based on the manufacturer: (1) air bearing or (2) magnetic bearing. In an air bearing turbo blower, an air film is formed between the impeller shaft and its bearings as the shaft rotates at high speed, achieving "friction free" floating of the shaft. Air bearing technology is offered by several manufacturers including K-Turbo, Neuros, Turblex, and HSI. In a magnetic bearing design, the impeller shaft is magnetically levitated to provide friction-free floating of the shaft. Turbo blowers featuring magnetic bearing design are offered by ABS Group, Atlas Copco, and Piller TSC. A magnetic bearing high-speed turbo blower is also being developed by Dresser Roots. The friction-free bearing design coupled with high efficiency motors contributes to the comparative high energy of the turbo blower technology. Turbo blowers have many practical advantages (Gass 2009; Jones and Burgess 2009):

- They are typically 10–20 percent more energy efficient than conventional multi-stage centrifugal or positive displacement for their current size range based on manufacturers' data. Good turndown capacity (up to 50 percent) with little drop in efficiency. It is important to note that efficiencies of turbo blowers at turndown are not yet well documented because the technology is so new.
- Some include a dynamic control package with integrated variable speed drive, sensors, and controls that automatically adjust blower output based on real-time dissolved oxygen (DO) demand in the aeration basin.
- They leave a small carbon footprint and are lightweight.

- They exhibit quiet operation, with low vibration, and sound enclosures are standard equipment used.
- They consist of few moving parts and require low maintenance.

Disadvantages of the turbo blower are that it is a new technology with relatively few installations, capital costs tend to be higher compared to other blower types, and multiple units may be needed for larger installations. Moreover, testing methods are not consistent among different manufacturers and some efficiency claims are not yet well documented.

SINGLE-STAGE CENTRIFUGAL BLOWERS

Single-stage centrifugal blowers equipped with inlet guide vanes pre-rotate the intake air before it enters the high-speed blower impellers. This reduces flow more efficiently than throttling. Blowers that are also equipped with variable outlet vane diffusers have improved control of the output air volume. Utilizing inlet guide vane and discharge diffusers on a single-stage centrifugal blower makes it possible to operate the blower at its highest efficiency point, not only at the design condition but also within a greater range outside of the design condition. Programmable logic controller (PLC) control can be used to optimize inlet guide vane operation (i.e., positioning) based on ambient temperature, differential pressure, and machine capacity. Automated DO and variable header pressure control can increase efficiency.

NEW DIFFUSER TECHNOLOGY

Public utility managers should be cognizant of the development of fine bubble diffuser technology in the 1970s that led to significant reductions in aeration energy consumption over mechanical and coarse bubble aeration due to the increased oxygen transfer rates afforded by the high surface area of the fine bubbler. Focus on Energy (2006) estimates that using fine bubble diffusion can reduce aeration energy by 25 percent to as high as 75 percent. Estimated energy savings of 30–40 percent are common (USEPA 1999a; Cantwell et. al., 2009).

There are many different types of fine bubble diffusers available, including ceramic/porous plates, tubular membranes, ceramic disks, ceramic domes, and elastomeric membrane disks, each with distinct advantages and disadvantages. In general, most diffusers are one of two types: (1) rigid ceramic material configured in discs or (2) perforated membrane material. Ceramic media diffusers have been in use for many years and are considered the standard against which new, innovative media are compared. Membrane diffusers consist of a flexible material with perforated pores through which air is released. Most often configured in tubes, discs or panels, they comprise the majority of new and retrofit installations.

Fine bubble aeration has been implemented at many WWTPs and is considered a common conventional ECM. The focus of this section is ECMs related to new diffuser equipment that can achieve enhanced energy reduction over fine bubble technology.

Recent advances in membrane materials have led to ultra-fine bubble diffusers, which generate bubbles with an average diameter of between 0.2 mm and 1.0 mm. The primary appeal of ultra-fine bubble diffusion is improved oxygen transfer efficiency (OTE). Additionally, some composite materials used in the manufacture of ultra-fine bubble diffusers are claimed to be more resistant to fouling, which serves to maintain the OTE and reduce the frequency of cleaning. Concerns about ultra-fine bubble diffusion include slow rise rates and the potential for inadequate mixing. Two proprietary ultra-fine bubble diffuser designs, panel diffusers by Parkson and Aerorstrip® diffusers by the Aerostrip Corporation, are discussed below.

Panel diffusers are membrane-type diffusers built onto a rectangular panel. They are designed to cover large areas of the basin floor and lay close to the floor. Panel diffusers are constructed of polyurethane and generate a bubble with a diameter of about 1 mm. OTE is a function of floor diffuser coverage, which translates to improved efficiency for panel diffusers. The advantages of panel diffusers include the increased OTE and the even distribution of aeration. Disadvantages include a higher capital cost, a higher head loss across the diffuser, increased air filtration requirements, and a tendency to tear when over-pressured.

Aerostrip® is a proprietary diffuser design manufactured in Austria by Aquaconsult. The device is a long strip diffuser with a large aspect ratio. According to the manufacturer,

> "it is a homogenous thermoplastic membrane held in place by a stainless steel plate." (Aerostrip)

The Aerostrip® diffuser provides many of the same advantages and disadvantages as panel diffusers; however, it appears to be less prone to tearing. Also, the smaller strips allow tapering of the diffuser placement to match oxygen demand across the basin. Aerostrips may be mounted at floor level or on supports above the floor. Manufacturer's claims regarding the strip membrane diffuser include:

- Energy efficiencies between 10 percent and 20 percent greater than the traditional ceramic and elastomeric membrane diffuser configurations
- Uniform bubble release across the membrane surface
- Bubbles resist coalescing
- Membranes not prone to clogging
- Diffusers are self-cleaning, although Aerostrip panels have been reported to be susceptible to frequent fouling requiring bumping and flexing of the membrane to dislodge (USEPA 2010)

INNOVATIVE AND EMERGING ENERGY CONSERVATION MEASURES

Unlike energy conservation measures for aeration and pumping, ECMs for advanced treatment technologies such as ultraviolet (UV) disinfection and membrane bioreactors (MBRs), and for other process functions, such as anoxic zone mixing, are emerging and generally not yet supported by operating data from full-scale installations. They are very important, however, because wastewater treatment plants (WWTPs) are increasingly employing these technologies. This section provides a discussion of ECMs for advanced technologies and presents full-scale plant test results where available. Where operation data are not available manufacturer's information is provided. This is the kind of information that public utility managers involved in wastewater need to be aware of.

UV Disinfection

Although ultraviolet disinfection was recognized as a method for achieving disinfection in the late 19th century, its application virtually disappeared with the evolution of chlorination technologies. However, in recent years, there has been resurgence in its use in the wastewater field, largely as a consequence of concerns related to security, safe handling, and effluent toxicity associated with chlorine residual. Even more recently, UV has gained more attention because of the tough new regulations on chlorine use imposed by both the Occupational Health and Safety Administration (OSHA) and the USEPA. Because of this relatively recent increased regulatory pressure, many facilities are actively engaged in substituting chlorine for other disinfection alternatives. Moreover, UV technology itself has made many improvements, which now makes UV attractive as a disinfection alternative. Loeng et al. (2008) estimated that as of 2007, approximately 21 percent of municipal WWTPs were using UV for disinfection. That number is only expected to rise as manufacturers continue to improve UV equipment designs and decrease costs, and as more and more WWTP gain experience with the technology.

UV radiation at certain wavelengths (generally 220–320 nm (nanometers)) can penetrate the cell walls of microorganisms and interfere with their genetic material. This limits the ability of microorganisms to reproduce and, thus, prevents them from infecting a host. UV radiation is generated by passing an electrical charge through mercury vapor inside a lamp. Low-pressure, low-intensity lamps, which are most common at WWTPs, produce mot radiation at 253.7 nm. Medium-pressure, high-intensity lamps emit radiation over a much wider spectrum and have 15 to 20 times the UV intensity of low-pressure, low-intensity lamps. Although fewer lamps are required as compared to low-pressure systems, medium-pressure lamps require more energy. The effectiveness of the UV process is dependent on:

- UV light intensity
- Contact time
- Wastewater quality (turbidity)

The Achilles' heel of UV for disinfecting wastewater is turbidity. If the wastewater quality is poor, the ultraviolet light will be unable to penetrate the solids and the effectiveness of the process decreases dramatically. For this reason, many states limit the use of UV disinfection to facilities that can reasonably be expected to produce an effluent containing ≤30 mg/L (milligrams per liter), or less of BOD_5 and total suspended solids.

In the operation of UV systems, UV lamps must be readily available when replacements are required. The best lamps are those with a stated operating life of at least 7,500 hours and those that do not produce significant amounts of ozone or hydrogen peroxide. The lamps must also meet technical specifications for intensity, output, and arc length. If the UV light tubes are submerged in the wastestream, they must be protected inside quartz tubes, which not only protect the lights but also make cleaning and replacement easier.

Contact tanks must be used with UV disinfection. They must be designed with the banks of UV lights in a horizontal position, either parallel or perpendicular to the flow or with banks of lights placed in a vertical position perpendicular to the flow.

Note: The contact tank must provide, at a minimum, 10-second exposure time.

It was stated earlier that turbidity problems have been one of the major problems with UV use in wastewater treatment—and this is the case. However, if turbidity is its Achilles' heel, then the need for increased maintenance (as compared to other disinfection alternatives) is the toe of the same foot. UV maintenance requires that the tubes be cleaned on a regular basis or as needed. In addition, periodic acid washing is also required to remove chemical buildup.

Routine operating of UV disinfection systems is required. Checking on bulb burnout, buildup of solids on quartz tubes, and UV light intensity is necessary (Spellman 2009).

Note: UV light is extremely hazardous to the eyes. Never enter an area where UV lights are in operation without proper eye protection. Never look directly into the ultraviolet light.

Energy requirements for UV depend on the number, type, and configuration of lamps used to achieve the target UV dose for pathogen inactivation. One of the most important factors affecting UV dose delivery is UV transmittance (UVT) of the water being disinfected. UVT is defined as the percent of light passing through a wastewater sample over a specific distance (1 cm (centimeter)). It takes into account the scattering and adsorption of UV by suspended and dissolved material it the water. UVT is affected by level of pretreatment—filtered wastewater has a much higher UVT than unfiltered water. Microorganisms that move quickly through the reactor far from the lamp will receive a lower dose than microorganisms that have longer exposure to the UV radiation and are closer to the lamp. Other factors affecting UV dose delivery are temperature, lamp age, and lamp fouling. Because UV disinfection is complex and based on many factors, dose estimation methods are complicated and typically involve computational fluid dynamic modeling or bioassays. Dose can be maintained at a minimum level or can be controlled based on water quality (i.e., lowered during periods of improved quality), which can save energy.

DID YOU KNOW?

The effectiveness of a UV disinfection system depends on the characteristics of the wastewater, the intensity of UV radiation, the amount of time the microorganisms are exposed to the radiation, and the reactor configuration. For any one treatment plant, disinfection success is directly related to the concentration of colloidal and particulate constituents in the wastewater (USEPA 1999b).

A study funded by the Pacific Gas and Electric (PG & E) company found that the energy consumed by UV disinfection can account for approximately 10–25 percent of total energy use at a municipal wastewater treatment plant (PG&E 2001). [**Note**: This is based on a detailed evaluation of seven wastewater treatment plants ranging in flow rate from 0.4 MGD to 43 MGD. Data set included plants with low-pressure, low-intensity lamps and higher-pressure high-intensity lamps in a variety of configurations]. Energy required for low-pressure lamps ranged from approximately 100–250 kWh/MG (kilowatt-hours per million gallons). Energy required for medium-pressure systems ranged from 460 kWh/MG to 560 kWh/MG, with one plant requiring 1,000 kWh/MG to achieve a very high level of coliform inactivation. PG&E (2001) reported that UV disinfection performance in relation to input energy is not linear. An increasing amount of energy is required to obtain marginal reductions in most probable number (MPN) per milliliter for total coliforms.

ECMs for UV disinfection are fairly new, and energy savings/cost data are not well documented in the literature. Still, growing experience with UV disinfection has revealed practical design, operation, and maintenance strategies that can reduce the energy use of UV disinfection. The following sections summarize these ECMs and provide detailed information on upgrades and associated energy savings for several WWTPs as reported in the literature.

Design

Pretreatment

Pretreatment to remove suspended solids from wastewater, such as tertiary sand filtration or membranes, can increase UV and allow a plant to reach the same level of treatment at a lower UV dose, thereby saving energy. If a plant uses iron or aluminum compounds for chemical precipitation of phosphorus, it is important to minimize residual iron and aluminum concentrations to prevent acceleration of UV lamp fouling (Loeng et al. 2008).

Lamp Selection

Medium-pressure lamps require two to four times more energy to operate than low-pressure, low-intensity lamps. In some cases, WWTPs can save on energy costs by specifying low-pressure, low-intensity lamps. Tradeoffs are (1) a larger footprint of the same disinfection level, which can be significant because as many as 20 low-pressure, low-intensity lamps are needed to produce the same disinfecting power as one medium-pressure lamp, and (2) higher operating costs for maintenance and change out of additional lamps.

Low-pressure, high-output lamps are similar to low-pressure, low-intensity lamps except that a mercury amalgam is used instead of mercury gas so they can operate at higher internal lamp pressures. Thus, the UV output of a low-pressure, high-output lamp is several times that of a low-pressure, low-output lamp (Loeng et al. 2008). Low-pressure, high-output lamps have significant advantages of:

- Reducing lamp requirements (i.e., quantity) compared to traditional low-intensity lamps
- Reducing energy requirements compared to medium-pressure lamps

Loeng et al. (2008) reported that the energy demand for low-pressure, high-output systems is similar to that of low-pressure, low-intensity systems. Thus, low-pressure, high-output lamps may be a good option for reducing the number of lamps and footprint while keeping the energy requirements low.

Salveson et al. (2009) presented results of a pilot test at the Stockton, CA, WWTP comparing design conditions and operation of medium-pressure and low-pressure, high-output lamps. The power draw for the low-pressure, high-output lamps was between 20 percent and 30 percent of the power draw [based on operational values (kW)] for the medium-pressure lamps, reducing annual O&M costs significantly. These results are similar to information reported from one manufacturer for a 30 MGD plant treating secondary effluent. A low-pressure, high-output system would use 60 kW at peak flow compared to 200 kW for a medium-pressure system (Faber 2010).

System Turndown

Similar to the design of blowers for aeration systems, it is important that designers allow for sufficient UV system turndown to respond to changes in flow and wastewater quality. Flexibility and control in design is a key factor in operating efficiently from the day the technology is commissioned until the end of its design life.

The configuration of the lamps dictates the approach for lamp turndown. In systems with vertical lamp configurations, the water level can vary during operation (with respect to the submerged portion of the lamp), whereas in a horizontal lamp configuration, the water levels should remain relatively constant (with respect to lamp submergence). Individual rows of lamps can be turned on and off in vertical configurations. In horizontal arrangements, channel control is more typically used to respond to varying flows (Loeng et al. 2008). Regardless of configuration, the number of channels should be selected to maintain a velocity that has been tested and is known to provide the required dose delivery.

System Hydraulics to Promote Mixing

As noted previously, UV dose delivery inside a UV reactor depends on the hydraulics. Optimized longitudinal and axial mixing of the water is critical to maintain a minimum UV dose throughout the reactor. In general, this is achieved by operating at a sufficiently high approach velocity to ensure turbulent flow conditions (WEF and ASCE 2010). WWTPs should conduct full-scale pilot testing before installation to ensure that mixing effects are addressed in design. Flow equalization prior to the UV reactor can also stabilize hydraulic conditions and prevent high or low flows from causing reduced UV disinfection performance. It is important to note that mixing is a balancing act. Extreme agitation of the wastewater can create bubble that shield pathogens from exposure to the UV radiation.

Operation and Maintenance

Automation

Automation can reduce the number of lamps and/or channels operating based on real-time flow and wastewater characteristic data, thereby reducing energy use and also extending UV lamp life. Controls can be designed to turn off lamps or divert flow to a few operating channels depending on the UV system design. Control is most commonly flow-paced control or dose-paced control. Flow-paced control is the simplest with a number of lamps/channels in service based strictly on influent flow rate. Dose-paced control is based on the calculated dose, which is derived from the following on-line monitoring data (Loeng et al. 2008):

- Flow rate
- UVT
- Lamp power (including lamp age and on-line intensity output data)

Dose-paced control more closely matches the UV dose delivered to wastewater conditions. For example, during periods of high solids removal, UVT will increase and UV output can be decreased to achieve the same dose. During wet weather events or other periods of low effluent quality, lamp output can be increased in response to reduced UVT.

At the University of California, Davis Wastewater Treatment Plant, process controls were implemented to divert flow automatically to one of two channels during low flow conditions (Phillips and Fan 2005). This change provided the flexibility to operate at 33, 50, 67 and 100 percent of maximum power. The original design limited operation to between 67 percent and 100 percent of maximum power. The annual energy use at the UC Davis WWTP is expected to decrease by 25 percent once the process changes are fully implemented.

Lamp Cleaning and Replacement

The effectiveness of UV disinfection systems depends on the intensity of the ultraviolet radiation to destroy the microorganism in the treated wastewater. Two factors that affect UV intensity during operation are lamp age and quartz sleeve fouling.

After an initial burn-in period, the lamp output decreases gradually toward the end of lamp life. The "end of lamp life" is defined by the manufacturer and is the operating hours at which the lamp reaches a specified minimum output. The operating life of UV lamps is provided below (WEF and ASCE 2010).

- Low-pressure, low-intensity lamps: 7,500–8,000 hours
- Low-pressure, high-intensity lamps: 12,000 hours
- Medium-pressure lamps: 5,000 hours

WWTPs can provide a relatively consistent level of lamp output by establishing a schedule for staging lamp replacements.

Algal growth, mineral deposits, and other materials can foul the lamp sleeve and subsequently decrease UV intensity and disinfection efficiency. Cleaning and maintaining quartz sleeves are critical to ensuring the optimum performance of UV disinfection and can result in substantial energy savings. Most equipment suppliers provide automatic cleaning mechanisms which consist of chemical cleaning, mechanical cleaning, or both. One study found that a combination of mechanical and chemical cleaning was superior to mechanical cleaning alone (Peng et al. 2005), cited in Loeng et al. 2008).

The Efficiency Partnership (2001) presents an example of energy savings due to increased attention to UV system cleaning and lamp replacement. At the Central Contra Costa Sanitary District (CCCSD), lamp cleaning and maintenance is particularly important because they are disinfecting secondary effluent with fairly low water quality. CCCSD found that increase maintenance of the UV lamps (i.e., the cleaning and replacement of UV bulbs) at its wastewater treatment plant resulted in a reduction in the number of UV banks required for the disinfection system from nine to six banks. Efficiency Partnership (2001) reported that this new maintenance strategy resulted in a power savings of 105 kW.

Membrane Bioreactors (MBRs)

Membrane bioreactors (MBRs) are becoming more common as WWTPs are required to meet increasingly stringent effluent limits and in some cases, reuse requirements in smaller footprints. The unique feature of MBRs is that instead of secondary clarification, they use membrane treatment, either as vacuum-driven systems immersed in a biological rector or as pressure-driven membrane systems located external to the bioreactor for solids separation. Membranes are typically configured hollow tube fibers or flat panels and have pore sizes ranging from 0.1 μ (microns) to 0.4 μ. Although MBRs have many operational advantages, they use more energy than conventional processes in order to move water through the membrane and for membrane scouring and cleaning. WEF (2009) reports that energy requirements of MBR systems may be twice that of conventional activated sludge systems.

Because the technology is not widespread, ECMs for MBRs are emerging. The emerging ECM identified in this report is membrane air scour alternatives.

Membrane fouling has been identified as the most significant technical challenge facing this technology (Ginzburg et al. 2008). Fouling occurs when the membrane pores become obstructed with the mixed liquor suspended solids being filtered, causing a loss in permeability. The main causes of membrane fouling are initial pore blocking where particles smaller than the membrane pore size plug the openings, followed by cake fouling, where particles accumulate on the membrane over time forming stratified "cake" layers (Peeters et al. 2007); Lim and Bai 2003). Although different membrane manufacturers use different techniques to control for fouling, the primary method to address cake fouling is aeration along with periodic chemical cleaning. Peeters et al. (2007) report that membrane aeration to control fouling accounts for 35–40 percent of total power consumption of an MBR.

In recent years, several membrane manufacturers have modified operational strategies to reduce air scour fouling control requirements (Wallis-Lage and Leversque 2009). For example, Kubota varies the volume of air used for aeration based on the flux (e.g., lower air scour rates are used for lower flux values). The manufacturer of the Huber system claims reduced energy consumption for air scour due to a centrally positioned air intake and lower pressure. Siemens uses a combination of air and water to scour the membrane (Wallis-Lage and Levesque 2009). General Electric (GE) implemented "cyclic" air scour whereby aeration would be turned on and off in 10-second intervals. A newer innovation is their 10/30 Eco-aeration where the membrane is scoured for 10 seconds on, 30 seconds off during non-peak flow conditions.

GE claims that the 10/30 Eco-aeration can reduce energy consumption by up to 50 percent compared to the standard 10/10 aeration protocol (Ginzburg et al. 2008).

The literature includes pilot- and full-scale test data for a membrane fouling controller and algorithm used to clean the GE ZENON ZeeWeed MBR. This system uses real-time analysis of the membrane's filtration operating conditions to determine the fouling mechanism present to the MBR system. The information obtained from the algorithm dictates the implementation of specific control actions to respond to the particular fouling mechanism (e.g., membrane aeration, backwash, chemical cleaning—the biggest impact on energy consumption being membrane aeration). When aeration is identified as the control action, the fouling controller/algorithm provides the MBR PLC system the information to select between the traditional 10/10 (air scour On/Off) protocol and a 10/30 Eco-Aeration energy-saving protocol. The algorithm was piloted and later full-scale tested at a 3 MGD plant in Pooler, Georgia (Ginzburg et al. 2008). Ginsburg (2008) concluded that additional research is required to further develop the on-line fouling controller to include additional control parameters such as membrane aeration flow rate, backwash flow rate, and backwash duration.

Anoxic and Anaerobic Zone Mixing

Many WWTPs are implementing biological nutrient removal (BNR) for nitrogen and/or phosphorus to protect receiving waters and prevent eutrophication, particularly in coastal regions. Biological nitrogen removal is a two-step process consisting of nitrification to convert ammonia to nitrate (NO_3) followed by denitrification to convert nitrate to nitrogen gas. Nitrification of ammonia is an aerobic process and can occur in the aerated zone with sufficient solids residence time (SRT). Significant energy can be required for complete nitrification of ammonia. Denitrification is an anoxic process accomplished in the absence of dissolved oxygen so that the microorganisms will use nitrate as their oxygen source. Denitrification can be accomplished in a denitrifying filter, but most often, it occurs in a suspended growth anoxic zone where the denitrifying microorganisms can use organic material present in the wastewater instead of or in addition to an external carbon source. A common configuration of the suspended growth nitrification-denitrification process is the Modified Ludzack-Ettinger (MLE) process, which has an initial anoxic zone followed by an aerobic zone. Nitrification occurs in the aerobic zone from which pumps recycle nitrate-rich mixed liquor to the anoxic zone for denitrification.

Biological phosphorus removal works by exposing the biomass first to anaerobic conditions. As long as a sufficient food source (i.e., volatile fatty acids) is present, microorganisms called phosphate accumulating organisms (PAOs) will release stored phosphorus in the anaerobic zone, which conditions them to uptake large amounts of phosphorus when they enter the aerobic zone. Phosphorus is removed when biomass is wasted from the aerobic zone.

It is important to mix the wastewater in the anoxic zone to maintain suspension of solids and ensure that denitrifying microorganisms come into contact with nitrate. Similarly, it is important to mix the wastewater in the anaerobic zone to maintain suspension of the solids and PAOs. The mixers, however, cannot impart oxygen to the water (this would cause them to use oxygen as their electron acceptor instead of nitrate). Similarly, for the anaerobic zone in biological phosphorus removal systems, mixers are needed to contact waste and microorganisms but must not transfer oxygen to the water (oxygen would promote growth of microorganisms other than PAOs which would compete with them for the food source). Low-speed submersible mixers are commonly used for these processes. Two emerging ECMs have been identified to reduce the energy required to mix anoxic and anaerobic zones: hyperbolic mixers and pulsed large bubble mixing.

Hyperbolic Mixing

A new hyperboloid mixer has undergone full-scale testing at two large wastewater treatment plants in the United States and has shown significant energy savings compared to traditional submersible mixers. The mixer is a vertical shaft-type mixer with a hyperboloid-shaped stirrer located close to the bottom of the tank. The stirrer is equipped with transport ribs that cause acceleration of the wastewater in a radial direction to promote complete mixing. The hyperboloid mixer has been used in Europe for more than 10 years with installations in Germany, Holland, and Belgium (Gidugu et al. 2010).

A recent study at the Bowery Bay water Pollution Control Plant (WPCP) in New York City compared the performance of traditional submersible mixers (specifically two blade propeller mixers mount on the side of the tank) to a hyperbolic mixer, the HYPERCLASSIC HC RKO 2500® (Fillos and Ramalingam 2005), for anoxic zone mixing. Researchers evaluated the two mixers based on their ability to:

- Sustain uniform distribution of suspended solids in the basin
- Maintain a low DO concentration (< 0.3 mg/L)
- Maintain a hydraulic profile supportive of denitrification (as determined using tracer tests)

Although both mixers at the Bowery Bay WPCP were able to achieve good distribution of solids with low DO, the Hyperclassic mixer had a superior hydraulic profile. Moreover, the authors reported lower energy needs for the Hyperclassic mixer due to its design: 2.2 bhp (brake horsepower) for the Hyperclassic mixer compared to 6.0 bhp for the submersible mixer. The authors reported a total energy cost of $1,131 for the Hyperclassic mixer compared to $3,075 for the submersible mixer per anoxic zone per year, for a savings of close to $2,000 based on a current energy rate of $0.039 per kWh. (**Note**: Based on this information, energy use for the Hyperclassic mixer would be 29,000 kWh/year compared to 78850 kWh/yr for the submersible mixer for an energy savings of 49,850 kWh/year per anoxic zone; Fillos and Ramalingam 2005.) The capital cost of the Hyperclassic mixer is approximately $10,000 more than the uniprop mixer, so simple payback would be approximately five years per anoxic zone.

Gidugu et al. (2010) reported results of side-by-side testing of the new hyperboloid mixer and a conventional hydrofoil mixer at the Blue Plains WWTP in Washington, D.C. The hydrofoil mixer, which is widely used in the United States, has a vertical shaft and a hydrofoil impeller with four angled stainless steel blades (Gidugu et al.

2010). Two 20 hp hydrofoil mixers were installed in one of the anoxic zones at the Blue Plains WWTP in October 2004 for evaluation. Six 10 hp hyperboloids mixers were installed in three anoxic zones (two per zone) for testing in October 2008. Researchers collected data to create DO and total suspended solids (TSS) profiles in all four anoxic zones in June 2008 to evaluate mixing.

Results showed similar results to Bowery Bay WPCP, with the hyperboloid mixer achieving good distribution of solids with low DO. TSS concentrations within the hyperboloid mixer were spread out over a smaller range of values than within the traditional hydrofoil mixer, indicating more uniform mixing. Gidugu et al. (2010) present a comparison of energy use, citing 9.7 bhp per unit for the 10 hp hyperboloid mixer compared to 17.3 bhp for the hydrofoil mixer. Based on an electricity cost of $0.10/kWh, they estimated energy savings potential of over $5,000 per year per mixer. At a cost difference of only $2,000 more for the hyperboloid mixer compared to the hydrofoil mixer, simple payback would be less than one year.

Pulsed Large Bubble Mixing

An innovative mixing technology by Enviromix called Bio^{M}x reduces energy required for anoxic of anaerobic zone mixing by firing short bursts of compressed air into the zone instead of mechanically mixing it. Uniquely designed nozzles produce a mass of large air bubbles, ranging from marble to softball sized, which mix the water as they rise to the surface (Randall and Randall 2010). The large air bubbles, much larger than those made by coarse bubble diffusers, are designed to minimize oxygen transfer and maintain anoxic or anaerobic conditions. The system includes a PLC to manage the timing of the air control valve firing, which gives the operator flexibility to respond to different conditions within the tank. The manufacturer reports that the system has non-clogging, self-cleaning in-tank components that require no maintenance.

An independent study at the F. Wayne Hill Water Resources Center in Gwinnett County, Georgia, compared the performance and energy use of Bio^{M}x to submersible propeller mixers (Randall and Randall 2010). The plant, treating 30 MGD on average with a design flow of 60 MGS, operates up to 10 parallel treatment trains each with anaerobic, anoxic, and aerobic zones for biological nitrogen and phosphorus removal. In the spring of 2009, the Bio^{M}x system was installed in two anaerobic cells of one treatment train. The system consisted of an Ingersoll Rand 5–15 hp variable speed rotary screw compressor, piping, controls, and floor mounted nozzles. Findings from the technology evaluation performed in January 2010 are summarized below.

- Dye tracer tests showed similar mixing for the Bio^{M}x and submersible mixer systems.
- Total suspended solids profiles showed that the Bio^{M}x unit is capable of mixing to homogeneity similarly to the submersible mixing units, although variability in the Bio^{M}x cells was slightly higher.
- Continuous oxidation-reduction potential (ORP) measurements over periods of 12 to 28 hours showed 95th percentile ORP values of less than -150 mv (millivolts), which is indicative of anaerobic environments. Given the

success in anaerobic environments (< -100 mv), the technology is also applicable for use in anoxic environments.
- Power analyzer readings taken simultaneously showed that energy (in kW) required to mix one anaerobic cell using the Bio^{M}x system was 45 percent less than the energy required by a submersible mixer. When operated in tree cells using the same compressor, 60 percent less energy was required (0.097 hp/1000 cf).

The manufacturer also presents test results conducted from April 2009 through February 2010, available online at http://enviro-mix.com/documents/FWayneHillEnergySuccess Story2009-091001.pdf.

REFERENCES AND RECOMMENDED READING

Brogdon, J., McEntyre, C., Whitehead, L., & Mitchell, J. 2008. *Enhancing the Energy Efficiency of Wastewater Aeration.* Presented at WEFTEC 2008. Chicago, IL. WEFTEC.

Burton, F.L. 1996. *Water and Wastewater Industries: Characteristics and Energy Management Opportunities.* Burton Environmental Engineering, Los Altos, CA: Prepared for the Electric Power Research Institute, Palo Alto, CA. Report CR106941.

Cantwell, J., Newton, J., Jenkins, T., Cavagnaro, P., & Kalwara, C. 2009. *Running an Energy Efficient Wastewater Utility Modifications that Can Improve Your Bottom Line.* Arlington, Virginia WEF Webcast.

Cormen, T.H., Leiserson, C.E., Rivest, R.L., & Stein, C. 2002. *Introduction to Algorithms.* 2nd ed. New Delhi: Prentice-Hall of India.

Efficiency Partnership. 2001. *Water/Wastewater Case Study: Central Contra Costa Sanitary District. Flex Your Power.* Available at http://fypower.org/pdf/CS_Water_CCCSD.pdf.

Efficiency Partnership. 2009. *Water/Wastewater Case Study: South Tahoe Public Utility District. Flex Your Power.* Available at http://fypower.org/pdf/CS_Water_Tahoe.pdf.

EPRI. 1998. *Quality Energy Efficiency Retrofits for Wastewater Systems.* Electrical Power Research Institute. Project Manager: Keith Carns. Washington, DC. CR-109081.

Faber, J. 2010. *In Evaluation of Energy Conservation Measures for Wastewater Treatment Facilities.* EPA 832-R-10-005 (2010). Washington, DC: United States Environmental Protection Agency.

Fillos, J., & Ramalingam, K. 2005. *Evaluation of Anoxic Zone Mixers at the Bowery Bay WPCP.* New York City Department of Environmental Protection, Process Planning Section, Division of Operations Support, Bureau of Wastewater Treatment. Contract # New York, NY PW-047.

Focus on Energy. 2006. *Water and Wastewater Energy Best Practice Guidebook.* Report Prepared by Science Applications International Corporation. Accessed 7/3/19 @ http://forcusoneenrgy.com/Businerss/Industrial-Business/Guidebooks.

Gass, J.V. (Black & Veatch). 2009. *Scoping the Energy Savings Opportunities in Municipal Wastewater Treatment.* Presented at the CEE Partner's Meeting, September 29, 2009. Available at http://cee1.org/cee/mtg/09-09mtg/fiels/WWWGAss.pdf.

Gidugu, S., Oton, S., & Ramalingam, K. 2010. Thorough Mixing Versus Energy consumption. *New England Water Environment Association Journal*, Spring.

Ginzburg, B., Peeters, J., & Pawloski, J. 2008. *On-Line Fouling Control for Energy Reduction in Membrane Bioreactors.* Presented at Membrane Technology, 2008. Atlanta, GA: WEF.

Gusfield, D. 1997. *Algorithms on Strings, Trees, and Sequences: Computer Science and Computational Biology.* Cambridge, UK, Cambridge University Press.

Jones, T., & Burgess, J. 2009. *Municipal Water-Wastewater Breakout Session: High Speed "Turbo" Blowers*. Presented at the Consortium for Energy Efficiency (CEE) Program Meeting, June 3, 2009. Available at http://cee1.org/cee/mtg/06-09mtg/files/WWW1Jo nesBurgess.pdf.

Lafore, R. 2002. *Data Structures and Algorithms in Java*. 2nd ed. New York, NY. SAMS.

Larson, L. 2009. *A Digital Control System for Optimal Oxygen Transfer Efficiency*. California Energy Commission, PIER Industrial/Agricultural/Water End-Use Energy Efficiency Program. Report CEC-500-2009-076. Available at http://energy.ca.gov/2009publicaito ns/CEC-500-2009-076/CEC-500-2009-076.PDF.

Lim, A.L., & Bai, R. 2003. Membrane Fouling and Cleaning in Microfiltration of Activated Sludge Wastewater. *Journal of Membrane Science*, 216, 279–290.

Littleton, H.X., Daigger, G.T., Amad, S., & Strom, P.F. 2009. *Develop Control Strategy to Maximize Nitrogen Removal and Minimize Operation Cost in Wastewater Treatment by Online Analyzer*. Presented at WEFTEC 2009. Orlando, FL: WEF.

Loeng, L.Y.C., Kup, J., & Tang, C. 2008. *Disinfection of Wastewater Effluent—Comparison of Alternative Technologies*. Alexandria, VA: Water Environment Research Foundation (SERF).

Love, N.G. 2000. *A Review and Needs Survey of Upset Early Warning Devices*. WERF Report 99-WWF-2. Alexandria, VA: WERG.

Lui, W., Lee, G., Schloth, P., & Serra, M. 2005. *Side by Side Comparison Demonstrated a 36 Percent Increase of Nitrogen Removal and 19 Percent of Aeration Requirements Using a Feed Froward Online Optimization System*. Presented at Arlington, VA. WEFTC 2005.

Mitchell, T.M. 1997. *Machine Learning*. New York: McGraw-Hill.

National Electrical Manufacturers Association. 2006. *NEMA Premium: Product Scope and Nominal Efficiency Levels, Including Tables 12-12 and 12-13 from NEMA Standards Publication MG 1-2006*. Accessed 2/3/19 @ http://nema.org/stds/conplimentary-docs/ upload/MG1premium.pdf.

Pacific Gas and Electric Company (PG&E). 2001. *Energy Benchmarking Secondary Wastewater Treatment and Ultraviolet Disinfection Processes at Various Municipal Wastewater Treatment Facilities*. San Francisco, CA. Available online at http://cee1.or /ind/mot-sys/ww/pge2.pdf.

Peeters, J., Pawloski, J., & Noble, J. 2007. *The Evolution of Immersed Hollow Fiber Membrane Aeration for MBR*. IWA 4th International Membrane Technologies Conference, Harrogate, UK, June 2007.

Peng, J., Qiu, Y., & Gehr, R. 2005. Characterization of Permanent Fouling on the Surfaces of UV Lamps Used for Wastewater Disinfection. *Water Environment Research*, 77(4), 309–322.

Phillips, D.L., & Fan, M.M. 2005. *Automated Channel Routing to Reduce Energy Use in Wastewater UV Disinfection Systems*. Davis, CA: University of California, Davis.

Poynton, C. 2003. *Digital Video and HDTV Algorithms and Interfaces*. Morgan Kaufmann, New York, NY.

Randall, C.W., & Randall, W.O. 2010. *Comparative Analysis of a Biomix System and a Submersible Propeller Mixer: Mixing in Anaerobic Zones at the f. Wayne Hill Water Resources Center, Buford, Georgia*. Report provided in Evaluation of Energy Conservation Measures (2010). Washington, DC: U.S. Environmental Protection Agency.

Salveson, A., Wade, T., Bircher, K., & Sotirakos, B. 2009. *High Energy Efficiency and Small Footprint with High-Wattage Low Pressure UV Disinfection for Water Reuse*. Presented at the International Ultraviolet Association (IUVA)/International Ozone Association (IGA) North American Conference, Boston, MA, June 2009.

Spellman, F.R. 2009. *Handbook of Water and Wastewater Treatment Plant Operations.* 2nd ed. Boca Raton, FL: CRC Press.
Spellman, F.R., & Drinan, J. 2001. *Fundamentals of Pumping.* Boca Raton, FL: CRC Press.
Tata, P., Patel, K., Soszynski, S, Lue-Hing, C., Carns, K., & Perkins, D. 2000. *Potential for the Use of On-Line Respirometry for the Control of Aeration.* Presented at WEFTEC, Anaheim, CA, October 2007.
Thumann, A., & Dunning, S. 2008. *Plant Engineers and Managers Guide to Energy Conservation.* 9th ed. New York: Fairmont Press.
Trillo, I., Jenkins, T., Redmon, D., Hilgart, T., & Trillo, J. 2004. *Implementation of Feedforward Aeration Control Using On-Line Offgas Analysis: The Grafton WWTP Experience.* Presented at WEFTEC 2004, New Orleans, LA, October 2004.
USDOE. 1996. *Replacing an Oversized and Underloaded Electric Motor.* Washington, DC: Office of Energy Efficiency and Renewable Energy, Industrial Technologies Program. Fact Sheet DOE/GO-10096-287. Accessed 2/3/19 @ http://eere.energy.gov/industry/bestpractices/pdfs/ms-2463.pdf.
USDOE. 2005a. *Motor Systems Tip Sheet #2: Estimating Motor Efficiency in the Field.* Industrial Technologies Program, Energy Efficiency and Renewable Energy. DOE/GO-102005-2021. Accessed 11/19/2019 @ http://eere.energy.gov/industry/bestpracites/pdfs/estimate_motor_efficiency_motor_systems2.pdf.
USDOE. 2005b. *Case Study—The Challenge: Improving Sewage Pump System Performance, Town of Trumbull.* U.S. Department of Energy, Energy Efficiency and Renewable Energy. Accessed 11/19/2019 @ http://eere.energy.gov/industry/bestpractes/case_study_sewage_pump.html.
USDOE. 2005c. *Onondaga County Department of Water Environment Protection: Process Optimization Saves Energy at Metropolitan Syracuse Wastewater Treatment Plant.* U.S. Department of Energy, Energy Efficiency and Renewable Energy. Accessed 11/19/2019 @ http://eere.energy.gov/industry/bestpractices/pdfs/onondaga_country.pdf.
USDOE. 2008. New Motor technologies Boost System efficiency. United States Department of Energy Industrial Technologies Program. *Energy Matters*, 211–232.
USEPA. 1999a. *Wastewater Technology Fact Sheet: Fine Bubble Aeration.* EPA 832-F-99-065. Available at http://epa.gov/OW-OWM.html/mtb/fin.pdf.
USEPA. 1999b. *Wastewater Technology Fact Sheet: Ultraviolet Disinfection.* Washington, DC: United States Environmental Protection Agency.
USEPA. 2003. *Technology Transfer Network Support Center for Regulatory Air Models.* Accessed at www.epa.gov/scram001/tt22.htm.
USEPA. 2008. *Ensuring a Sustainable Future: An Energy Management Guidebook for Wastewater and Water Utilities.* Washington, DC: United States Environmental Protection Agency.
USEPA. 2010. *Evaluation of Energy Conservation Measures for Wastewater Treatment Facilities.* EOA 832-R-10-005. Washington, DC: Environmental Protection Agency.
Wagner, M., & von Hoessle, R. 2004. Biological Coating of EPDM-Membranes of Fine Bubble Diffusers. *Water Science and Technology*, 50(7), 79–85.
Wallis-Lage, C.L., & Levesque, S.D. 2009. *Cost Effective & Energy Efficient MBR Systems.* Presented at the Singapore International Water Week, June 22–26, 2009. Suntec Singapore International Convention and Exhibition Center. Available at http://bywater.files.wordpress.com/2009/05/abstact_wiw09_wallis-lage.pdf.
Water Environment Federation (WEF). 2009. *Manual of Practice (MOP) No. 32: Energy Conservation in Water and Wastewater Facilities.* New York: McGraw-Hill.
Weerapperuma, D., & de Silva, B. 2004. *On-Line Analyzer Applications for BNR Control.* Presented at WEFTEC, New Orleans, LA.

WEF. 2009. *MOP No. 32: Energy Conservation in Water and Wastewater Facilities*. Prepared by the Energy Conservation in Water and Wastewater Treatment Facilities Task Force of the Water Environment Federation. New York: McGraw-Hill.

WEF and ASCE. 2006. *Biological Nutrient Removal (BNR) Operation in Wastewater Treatment Plants—MOP 29*. Water Environment Federation and the American Society of Civil Engineers. Alexandria, VA: WEF Press.

WEF and ASCE. 2010. *Design of Municipal Wastewater Treatment Plants—WEF Manual of Practice 8 and ASCE Manuals and Reports on Engineering Practice NO. 76*. 5th ed. Alexandria, VA: Water Environment Federation/Reston, VA: American Society of Civil Engineers Environment & Water Resources Institute.

8 Digital Network Security

We depend on critical infrastructure every day. Our ability to travel, to communicate with friends and family, to conduct business, to handle our finances, and even our ability to access clean, safe food and water are all reliant upon our Nation's critical infrastructure networks and systems. These essential services that underlie daily life in American society are increasingly being run on digital networks. Every day, people connect to the national grid without even realizing it from their smart phones, computers, and tablets. As a result, these critical systems are prime targets for cyber attacks from those seeking to cause our country harm.

DHS (2016)

Note: In this book we use the terms information technology (IT), cyber space and digital network interchangeably.

INTRODUCTION

On April 23, 2000, police in Queensland, Australia, stopped a car on the road and found a stolen computer and radio inside. Using commercially available technology, a disgruntled former employee had turned his vehicle into a pirate command center of sewage treatment along Australia's Sunshine Coast. The former employee's arrest solved a mystery that had troubled the Maroochy Shire wastewater system for two months. Somehow the system was leaking hundreds of thousands of gallons of putrid sewage into parks, rivers and the manicured grounds of a Hyatt Regency hotel—marine life died, the creek water turned black and the stench was unbearable for residents. Until the former employee's capture—during his 46th successful intrusion—the utility's managers did not know why.

Specialists study this case of cyber-terrorism because, at the time, it was the only one known in which someone used a digital control system deliberately to cause harm. The former employee's intrusion shows how easy it is to break in—and how restrained he was with his power.

To sabotage the system, the former employee set the software on his laptop to identify itself as a pumping station, and then suppressed all alarms. The former employee was the "central control station" during his intrusions, with unlimited command of 300 SCADA nodes governing sewage and drinking water alike.

127

The bottom line: as serious as the former employee's intrusions were they pale in comparison with what he could have done to the fresh water system—he could have done anything he liked.

Barton Gellman 2002

Other reports of digital network exploits illustrate the debilitating effects such attacks can have on the nation's security, economy, and on public health and safety.

- In May 2015, media sources reported that data belonging to 1.1 million health insurance customers in the Washington, D.C., area were stolen in a cyber attack on a private insurance company. Attackers accessed a database containing names, birth dates, email addresses, and subscriber ID numbers of customers.
- In December 2014, the Industrial Control Systems Cyber Emergency Response Team (ICS-CERT, which works to reduce risks within and across all critical infrastructure sectors by partnering with law enforce agencies) issued an updated alert on a sophisticated malware campaign compromising numerous industrial control system environments. Their analysis indicated that this campaign had been ongoing since at least 2011.
- In the January 2014 to April 2014 release of its Monitor Report, ICS-CERT reported that a public utility had been compromised when a sophisticated threat actor gained unauthorized access to its control system network through a vulnerable remote access capability configured on the system. The incident highlighted the need to evaluate security controls employed at the perimeter and ensure that potential intrusion vectors are configured with appropriate security controls, and monitoring and detection capabilities.
- In December 2016, a Wisconsin couple was charged after duo allegedly defrauded Enterprise Credit Union in Brookfield of more than $300,000 after one of the defendants, who managed the bank's accounts, had her co-conspirator encash bank checks worth $980 several times each week beginning in May 2015. The charges allege that the couple used the money to buy drugs (Source: http://wauwatrosanow.com/story/news/cirme/2016/12/09/two-charged-allegedly-scamming-credit-unions-over-300k/95207718/.).

In 2000, the FBI identified and listed threats to critical infrastructure. These threats are listed and described in Table 8.1. In 2015, the Government Accountability Office (GAO) described the sources of digital network-based threats. These threats are listed and described in detail in Table 8.2.

DID YOU KNOW?

Presidential Policy Directive 21 defined "All hazards" as a threat to an incident, natural or man-made, that warrants action to protect life, property, the environment, and public health or safety, and to minimize disruptions of government, social, or economic activities.

TABLE 8.1
Threats to Critical Infrastructure Observed by the FBI

Threat	Description
Criminal groups	There is an increased use of cyber intrusions by criminal group who attack systems for purposes of monetary gain.
Foreign intelligence services	Foreign intelligence services use cyber tools as part of their information-gathering and espionage activities.
Hackers	Hackers sometimes crack into networks for the thrill of the challenge or for bragging rights in the hacker community. While remote cracking once required a fair amount of skill or computer knowledge, hackers can now download attack scripts and protocols from the Internet and launch them against victim sites. Thus, while attack tools have become more sophisticated, they have also become easier to use.
Hacktivists	Hacktivism refers to politically motivated attacks on publicly accessible Web pages or email servers. These groups and individuals overload email servers and hack into Web sites to send a political message.
Information warfare	Several nations are aggressively working to develop information warfare doctrine, programs, and capabilities. Such capabilities enable a single entity to have a significant and serious impact by disrupting the supply, communications, and economic infrastructures that support military power—impacts that, according to the Director of Central Intelligence, can affect the daily lives of Americans across the country.
Inside threat	The disgruntled organization insider is a principal source of computer crimes. Insiders may not need a great deal of knowledge about computer intrusions because their knowledge of a victim system often allows them to gain unrestricted access to cause damage to the system or to steal system data. The insider threat also includes outsourcing vendors.
Virus writers	Virus writers are posing an increasingly serious threat. Several destructive computer viruses and "worms" have harmed files and hard drives, including the Melissa Macro Virus, the Explore.Zip worm, the CIH (Chernobyl) Virus, Nimda, and Code Red.

Source: FBI, 2000; 2014.

Threats to systems supporting critical infrastructure are evolving and growing. As shown in Table 8.2, cyber threats can be unintentional or intentional. Unintentional or non-adversarial threats include equipment failures, software coding errors, and the actions of poorly trained employees. They also include natural disasters and failures of critical infrastructure on which the organization depends but which are outside of its control. Intentional threats include both targeted and untargeted attacks form a variety of sources, including criminal groups, hackers, disgruntled employees, foreign nation engaged in espionage and information warfare, and terrorists. These threat adversaries vary in terms of the capabilities of the actors, their willingness to act, and their motives, which can include seeking monetary gain or seeking an economic, political, or military advantage (GAO 2015).

TABLE 8.2
Common Digital Network Threat Sources

Source	Description
Non-adversarial, malicious	
Failure in information technology equipment	Failures in displays, sensors, controllers, and information technology hardware responsible for data storage, processing, and communications
Failure in environmental controls	Failures in temperature/humidity controllers or power supplies
Software coding errors	Failures in operating systems, networking, and general-purpose and mission-specific applications
Natural or man-made disaster	Events beyond an entity's control such as fires, floods/tsunamis, tornadoes, hurricanes, and earthquakes
Unusual or natural event	Natural events beyond the entity's control that are not considered to be disasters (e.g., sunspots)
Infrastructure failure or outage	Failure or outage of telecommunications or electrical power
Unintentional user errors	Failures resulting from erroneous, accidental actions taken by individuals (both system users and administrators) in the course of executing their everyday responsibilities
Adversarial	
Hackers or hacktivists	Hackers break networks for the challenge or for revenge, stalking, or monetary gain, among other reasons. Hactivists are ideologically motivated actors who use cyber exploits to further political goals.
Malicious insiders	Insiders (e.g., disgruntled organization employees, including contractors) may not need a great deal of knowledge about computer intrusions because their position with the organization often allows them to gain unrestricted access and cause damage to the target system or to steal system data. These individuals engage in purely malicious activities which should not be confused with non-malicious insider accidents.
Nations	Nations, as well as nation-states, and state-sponsored and state-sanctioned programs use cyber tools as part of their information-gathering and espionage activities. In addition, several nations are aggressively working to develop information warfare doctrine, programs, and capabilities.
Criminal groups and organize crime	Criminal groups seek to attack systems for monetary gain. Specifically, organized criminal groups use cyber exploits to commit identity theft, online fraud, and computer extortion.
Terrorists	Terrorists seek to destroy, incapacitate, or exploit critical infrastructures in order to threaten national security, cause mass casualties, weaken the economy, and damage public morale and confidence.
Unknown malicious outsiders	Unknown malicious outsiders are threat sources or agents that, due to a lack of information, agencies are unable to classify as being one of the five types of threat sources or agents listed above.

Source: GAO analysis of unclassified government and nongovernmental data. GAO 16–79.

THE DIGITAL NETWORK

Today's developing "information age" technology has intensified the importance of critical infrastructure protection, in which digital network or cyber security has become as critical as physical security to protecting virtually all critical infrastructure sectors. The Department of Defense (DoD) determined that cyber threats to contractors' unclassified information systems represented an unacceptable risk of compromise to DoD information and posed a significant risk to U.S. national security and economic security interests.

In the past few years, especially since 9/11, it has been somewhat routine for us to pick up a newspaper, magazine, or view a television news program where a major topic of discussion is cyber security or its lack thereof. For example, recently there has been discussion about Russian hackers trying to influence the U.S. 2016 elections. Many of the cyber intrusion incidents we read or hear about have added new terms or new uses for old terms to our vocabulary. For example, old terms such as botnets (short for robot networks, also balled bots, zombies, botnet fleets, and many others) which are groups of computers that have been compromised with malware such as Trojan horses, worms, backdoors, remote control software, and viruses have taken on new connotations in regard to cyber security issues. Relatively new terms such as scanners, Windows NT hacking tools, ICQ hacking tools, mail bombs, sniffer, logic bomb, nukers, dots, backdoor Trojan, key loggers, hackers' Swiss knife, password crackers, blended threats, Warhol worms, flash threats, targeted attacks, and BIOS crackers are now commonly read or heard. New terms have evolved along with various control mechanisms. For example, because many control systems are vulnerable to attacks of varying degrees, these attack attempts range from telephone line sweeps (wardialing), to wireless network sniffing (wardriving), to physical network port scanning, to physical monitoring and intrusion. When wireless network sniffing is performed at (or near) the target point by a pedestrian (warwalking), meaning that instead of a person being in an automotive vehicle, the potential intruder sniffing the network for weaknesses or vulnerabilities may be on foot, posing as a person walking, but they may have a handheld personal digital assistant (PDA) device or laptop computer (Warwalking 2003). Further, adversaries can leverage common computer software programs, such as Adobe Acrobat and Microsoft Office, to deliver a threat by embedding exploits within software files that can be activated when a user opens a file within its corresponding program. Finally, the communications infrastructure and the utilities are extremely dependent on the information technology sector. This dependency is due to the reliance of the communications systems on the software that runs the control mechanism of the operations systems, the management software, the billing software, and any number of other software packages is used by industry. Table 8.3 provides descriptions of common exploits or techniques, tactics, and practices used by digital/cyber adversaries.

Not all relatively new and universally recognizable information technology and digital or cyber terms have sinister connotations or meaning, of course. Consider, for example, the following digital terms: backup, binary, bit byte, CD-ROM, CPU, database, email, HTML, icon, memory, cyberspace, modem, monitor, network, RAM,

TABLE 8.3
Common Methods of Digital/Cyber Exploits

Exploit	Description
Watering hole	A method by which threat actors exploit the vulnerabilities of carefully selected websites frequented by users of the targeted system. Malware is then injected to the targeted system via the compromised websites
Phishing and spear phishing	A digital form of social engineering that uses authentic-looking emails, websites, or instant messages to get users to download malware, open malicious attachments, or open links that direct then to a website that requires information or executes malicious code
Insufficient authentication requirements	An exploit that takes advantage of a system's insufficient user authentication and/or any elements of cyber security supporting it to include not limiting the number of failed login attempts, the use of hard-coded credentials, and the use of a broken or risky cryptographic algorithm
Trusted third parties	An exploit that takes advantage of the security vulnerabilities of trusted third parties to gain access to an otherwise secure system
Classic buffer overflow	An exploit that involves the intentional transmission of more data than a program's input buffer can hold, leading to the deletion of critical data and subsequent execution of malicious code
Cryptographic weakness	An exploit that takes advantage of a network employing insufficient encryption when either storing or transmitting data, enabling adversaries to read and/or modify the data stream
Structured Query Language (SQL) injection	An exploit that involves the alteration of a database search in a web-based application, which can be used to obtain unauthorized across to sensitive information in a database, resulting in data loss at corruption, denial or service, or complete host takeover
Operating system command injection	An exploit that takes advantage of a system's inability to properly neutralize special elements used in operating system commands, allowing the adversaries to execute unexpected commands on the system by either modifying already evoked commands or evoking their own.
Cross-site scripting	An exploit that uses third-party web resources to run lines of programming code (referred to as scripts) within the victim's web browser or scriptable application. This occurs when a user, using a browser, visits a malicious website or clicks a malicious link. The most dangerous consequences can occur when this method is used to exploit additional vulnerabilities that may permit an adversary to steal cookers (data exchanged between a web server and a browser), log key stokes, capture screen shots, discover and collect network information, or remotely access and control the victim's machine.
Cross-site request forgery	An exploit takes advantage of an application that cannot, or does not, sufficiently verify whether a well-formed, valid, consistent request was intentionally provided by the user who submitted the request, tricking the victim into executing a falsified requires that results in the system or data being compromised.
Path traversal	An exploit that seeks to gain access to files outside of a restricted directory by modifying the directory pathname in an application that does not properly neutralize special elements (e.g., "…," "/," "…/," etc.)

(Continued)

TABLE 8.3 (CONTINUED)
Common Methods of Digital/Cyber Exploits

Exploit	Description
Integer overflow	An exploit where malicious code is inserted that leads to unexpected integer overflow, or wraparound, which can be used by adversaries to control looping or make security decisions in order to cause program crashes, memory corruption, or the execution of arbitrary code via buffer overflow
Uncontrolled format string	Adversaries manipulate externally controlled format strings in print-style functions to gain access to information and/or execute unauthorized code or commands.
Open redirect	An exploit where the victim is tricked into selecting a URL (website location) that has been modified t direct them to an external, malicious site which may contain malware that can compromises the victim's machine
Heap-based buffer overflow	Similar to classic buffer overflow, but the buffer that is overwritten is allocated in the heap portion of memory, generally meaning that the buffer was allocated using a memory allocation routine, such as "malloc ( )".
Unrestricted upload of files	An exploit that takes advantage of insufficient upload restrictions, enabling adversaries to upload malware (e.g., .php) in place of the intended file type (e.g., .jpg)
Inclusion of functionality from an un-trusted sphere	An exploit that uses trusted, third-party executable functionality (e.g., web widget or library) as a means of executing malicious code in software whose protection mechanism is unable to determine whether the functionality is from a trusted source, modified in transit, or being spoofed
Certificate and certificate authority compromise	Exploits facilitated via the issuance of fraudulent digital certificates (e.g., transport layer security and Secure Socket Layer). Adversaries use these certificates to establish secure connections with the target organization or individual by mimicking a trusted third party
Hybrid of others	An exploit which combines elements of two or more of the aforementioned techniques

Source: GAO, (2015).

Wi-Fi (wireless fidelity), record, software, and World Wide Web—none of these terms normally generate thoughts of terrorism in most of us.

THE BOTTOM LINE

United States Department of Homeland Security, in collaboration with many stakeholders of the information technology sector, identified cyber risk as significant to all critical infrastructure sectors. In the next chapter, as an example, we detail SCADA and its importance, significances, and vulnerabilities, all of which point to the need for planning, preparation, and mitigation procedures to ensure the resilience of digital network systems and applications.

REFERENCES AND RECOMMENDED READING

Associated Press (AP). 2009. Goal: Disrupt. *The Virginian-Pilot*, Norfolk, VA, 04 April 2009.
DHS. 2009. *National Infrastructure Protection Plan*. Retrieved 05 November 2017 @ http://dhs.gov/xlibrary/assets/NIPP.Plan.pdf.
DHS. 2003. *The National Strategy for the Physical Protection of Critical Infrastructures and Key Assets*. Accessed 12 December 2019@ https://dhs.gov/xlibrary/assets/physical_Strat.
DHS. 2013. *Homeland Security Directive 7: Critical Infrastructure Identification, Prioritization, and Protection*. Accessed 12 December 2019 @ https://dihs.gov/xabout/laws/gc_1214597989952.shtm.
DOE. 2001. *21 Steps to Improve Cyber Security of SCADA Networks*. Washington, DC: Department of Energy.
FEMA. 2008. *FEMA452: Risk Assessment A How to Guide*. Accessed 05 January 2019 @ fema.gov/library/file?type=published/filetofile.
FEMA. 2015. *Protecting Critical Infrastructure Against Insider Threats*. Accessed 17 April 2019 @ http://emilms.fema.gov/IS0915/IABsummary.htm.
FBI. 2000. *Threat to Critical Infrastructure*. Washington, DC: Federal Bureau of Investigation.
FBI. 2007. *Ninth Annual Computer Crime and Security Survey*. FBI: Computer Crime Institute and Federal Bureau of Investigations, Washington, DC.
FBI. 2014. *Protecting Critical Infrastructure and the Importance of Partnerships*. Accessed 12 December 2019 @ https://fib.gov/news/speeches/protecitng-critical-infrastruce-and t-the-importantce-o.
GAO. 2003. *Critical Infrastructure Protection: Challenges in Securing Control System*. Washington, DC: United States General Accounting Office.
GAO. 2015. *Critical Infrastructure Protection: Sector-Specific Agencies Need to Better Measure Cybersecurity Progress*. Washington, DC: United States Government Accountability Office.
Gellman, B. 2002. Cyber-Attacks by Al Qaeda Feared: Terrorists at Threshold of Using Internet as Tool of Bloodshed, Experts Say. *Washington Post*, June 27, p. A01.
Minter, J.G. 1996. Prevention Chemical Accidents Still a Challenge. *Occupational Hazards*, September, xix, 66–69.
National Infrastructure Advisory Council. 2008. *First Report and Recommendations on the Insider Threat to Critical Infrastructure*. Department of Homeland Security, Washington, DC.
NIPC. 2002. *National Infrastructure Protection Center Report*. Washington, DC: National Infrastructure Protection Center.
Spellman, F.R. 1997. *A Guide to Compliance for PSM/RMP*. Lancaster, PA: Technomic Publishing Company.
Stamp, J., et al. 2003. *Common Vulnerabilities in Critical Infrastructure Control Systems*, 2nd ed. Sandia National Laboratories.
US DOE. 2002. *Vulnerability Assessment Methodology: Electric Power Infrastructure*. Department of Energy, Washington, DC.
USEPA. 2005. EPA Needs to Determine What Barriers Prevent Water Systems from Securing Known SCADA Vulnerabilities. In: Harris, J. (ed.), *Final Briefing Report*. Washington, DC: USEPA, p.11–17.
Warwalking. 2003. Warwalking, Accessed 05 September 19 @ http://warwalking.tribe.net.

9 SCADA

Nobody has ever been killed by a cyberterrorist.

Unless people are injured, there is less drama and emotional appeal.

Dorothy Denning

WHAT IS SCADA?

Note: SCADA was briefly mentioned earlier in the text but because it is so critical to modern public utility operations, it is presented in the following in greater detail.

The critical infrastructure of many countries is increasingly dependent on SCADA systems.

If we were to ask the specialist to define SCADA, the technical response could be outlined as follows:

- SCADA is a multi-tier system or an interface with multi-tier systems.
- It is used for physical measurement and to control endpoints via a Remote Terminal Unit (RTU) and Program Logic Controller (PLC) to measure voltage, adjust a value, or flip a switch.
- It is an intermediate processor normally based on commercial third-party OSes—VMS, Unix, Windows, Linux, etc.
- It is used in human interfaces, for example, with graphical user interface (Windows graphic user interface (GUIs)).
- Its communication infrastructure consists of a variety of transport mediums such as analog, serial, Internet, radio, and Wi-Fi.

How about the non-specialist response, for the rest of us who are non-specialists? Well, for those of us in this category we could simply say that SCADA is a computer-based control system that remotely controls and monitors processes previously controlled and monitored manually. The philosophy behind SCADA control systems can be summed up by the phrase "If you can measure it, you can control it." SCADA allows an operator using a central computer to supervise (control and monitor) multiple networked computers at remote locations. Each remote computer can control mechanical processes (mixers, pumps, valves, etc.) and collect data from sensors at its remote location. Hence the name "Supervisory Control and Data Acquisition."

The central computer is called the Master Terminal Unit (MTU). The MTU has two main functions: periodically obtaining data from RTUs/PLCs and controlling remote devices through the operator station. The operator interfaces with the MTU using software called Human Machine Interface (HMI). The remote computer is called Program Logic Controller or Remote Terminal Unit. The RTU activates a relay (or switch) that turns mechanical equipment "on" and "off." The RTU also collects data from sensors. Sensors perform measurement, and actuators perform control.

In the initial stages utilities ran wires, also known as hardwire or land lines, from the central computer (MTU) to the remote computers (RTUs). Because remote locations were located hundreds of miles from the central location, utilities began to use public phone lines and modems, leased telephone company lines, and radio and microwave communication. More recently, they have also begun to use satellite links, Internet, and newly developed wireless technologies.

DID YOU KNOW?

Modern RTUs typically use a ladder-logic approach to programming due to its similarity to standard electrical circuits. An RTU that employs this ladder-logic programming is called a Programmable Logic Controller.

Because SCADA systems' sensors provided valuable information, many critical infrastructure entities, utilities, and other industries established "connections" between their SCADA systems and their business system. This allowed utility/industrial management and other staff access to valuable statistics, such as chemical usage. When utilities/industries later connected their systems to the Internet, they were able to provide stakeholders/stockholders with usage statistics on the IT segment and utility/industrial web pages. Figure 9.1 provides a basic illustration of a representative SCADA network. Note that firewall protection would normally be placed between Internet and business systems and between business systems and the MTU.

SCADA APPLICATIONS

As stated above, SCADA systems can be designed to measure a variety of equipment operating conditions and parameters or volumes and flow rates or electricity, natural gas and oil or oil and petrochemical mixture quality parameters, and to respond to change in those parameters either by alerting operators or by modifying system operation through a feedback loop system without having personnel physically visit each valve, process, or piece of other equipment on a daily basis to check it and/or ensure that it is functioning properly. Automation and integration of large-scale diverse assets required SCADA systems to provide the utmost in flexibility, scalability, openness, and reliability. SCADA systems are used to automate certain energy production functions; these can be performed without initiation by an operator. In addition to process equipment, SCADA systems can also integrate specific security alarms and equipment, such as cameras, motion sensors, lights, data from card reading systems, etc., thereby providing a clear picture of what is happening in all areas throughout a facility. Finally, SCADA systems also provide constant, real-time data on processes, equipment, location access, etc., enabling the necessary response to be made quickly. This can be extremely useful during emergency conditions, such as when energy distribution lines or piping breaks or when potentially disruptive

SCADA

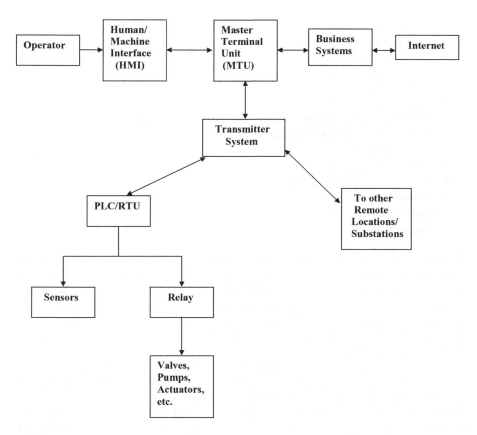

FIGURE 9.1 Representative SCADA network.

chemical reaction spikes appear in chemical processing operations. Currently, it can be said that SCADA has evolved from a simple indicating light and pushbutton control system into a comprehensive operation and handling system for very complex process control and safety shutdown systems. In a nutshell, SCADA results in an oversight system that requires fewer operators.

Today, many common digital system applications for SCADA systems include, but are not limited to, those given below.

- Boiler controls
- Bearing temperature monitors (electric generators and motors)
- Gas processing
- Plant monitoring
- Plant energy management
- Power distribution monitoring
- Electric power monitoring
- Fuel oil handling system
- Hydroelectric load management

- Petroleum pilot plants
- Plant monitoring
- Process controls
- Process stimulators
- Tank controls
- Utility monitoring
- Safety parameter display systems and shutdown systems
- Tank level control and monitoring
- Turbine controls
- Turbine monitoring
- Virtual annunciator panels
- Alarm systems
- Security equipment
- Event logging

Because these systems can monitor multiple processes, equipment, and infrastructure and then provide quick notification of, or response to, problems or upsets, SCADA systems typically provide the first line of detection for atypical or abnormal conditions. For example, a SCADA system connected to sensors that measure specific machining quality parameters are measured outside of a specific range. A real-time, customized operator interface screen could display and control critical systems monitoring parameters.

The system could transmit warning signals back to the operators, such as by initiating a call to a personal pager. This might allow the operators to initiate actions to prevent power outages or contamination and disruption of the energy supply. Further automation of the system could ensure that the system initiated measures to rectify the problem. Preprogrammed control functions (e.g., shutting a valve, controlling flow, throwing a switch, or adding chemicals) can be triggered and operated based on SCADA utility.

SCADA VULNERABILITIES

U.S. Electric Grid Gets Hacked Into

The Associated Press (AP) reported on April 9, 2009, that spies hacked into the U.S. energy grid and left behind computer programs (Trojan horses) that would enable them to disrupt service, exposing potentially catastrophic vulnerabilities in key pieces of national infrastructure.

Even though terrorists, domestic and/or foreign, tend to aim their main focus around the critical devices that control actual critical infrastructure production and delivery activities, according to USEPA (2005), SCADA networks were developed with little attention paid to security, often making the security of these systems weak. Studies have found that, while technological advancements introduced vulnerabilities, many critical infrastructure sector plans/sites and utilities have spent little time securing their SCADA networks. As a result, many SCADA networks may be susceptible

SCADA

to attacks and misuse. SCADA systems languished in obscurity, and this was the essence of their security—that is, until technological developments transformed SCADA from a backroom operation to a front-and-center visible control system.

Remote monitoring and supervisory control of processes began to develop in the early 1960s, and adopted many technological advancements. The advent of minicomputers made it possible to automate a vast number of once manually operated switches. Advancements in radio technology reduced the communication costs associated with installing and maintaining buried cable in remote areas. SCADA systems continued to adopt new communication methods, including satellite and cellular communication. As the price of computers and communications dropped, it became economically feasible to distribute operations and to expand SCADA networks to include even smaller facilities.

Advances in information technology and the necessity of improved efficiency have resulted in increasingly automated and interlinked infrastructures, and created new vulnerabilities due to equipment failure, human error, weather and other natural causes, and physical and cyber attacks. Some areas and examples of possible SCADA vulnerabilities include (Wiles, et al., 2007):

- Human—people can be tricked or corrupted, and may commit errors
- Communications—messages can be fabricated, intercepted, changed, deleted, or blocked
- Hardware—security features are not easily adapted to small self-contained units with limited power supplies
- Physical—intruders can break into a facility to steal or damage SCADA equipment
- Natural—tornadoes, floods, earthquakes, and other natural disasters can damage equipment and connections
- Software—programs can be poorly written

Specific SCADA weaknesses and potential attack vectors include:

- Absence of authorization requirements
- Absence of encryption
- Improper recognition and handling of errors and exceptions
- Absence of authentication requirements
- Possibility of data interception
- Possibility of manipulation of data
- Possibility of service denial
- Possibility of IP address spoofing—Internet protocol packets with a false source IP address
- Danger of session hijacking
- Unsolicited responses
- Scope for packet fuzzing—involves imputing false information, etc.
- Unauthorized control
- Log data medication

SCADA system computers and their connections are susceptible to different types of information system attacks and misuse such as those mentioned above. The Computer Security Institute and Federal Bureau of Investigation conduct an annual Computer Crime and Security Survey (FBI, 2007). The survey reported on 10 types of attacks or misuse, and reported that virus and denial of service had the greatest negative economic impact. The same study also found that 15 percent of the respondents reported abuse of wireless networks, which can be a SCADA component. On average, respondents from all sectors did not believe that their organization invested enough in security awareness. For example, utilities as a group reported a lower average computer security expenditure/investment per employee than many other sectors such as transportation, telecommunications, and finance.

Sandia National Laboratories' *Common Vulnerabilities in Critical Infrastructure Control Systems* described some of the common problems it has identified in the following five categories (Stamp et al., 2003):

1. **System Data**—important data attributes for security include availability, authenticity, integrity, and confidentiality. Data should be categorized according to its sensitivity, and ownership and responsibility must be assigned accordingly. However, SCADA data is often not classified at all, making it difficult to identify where security precautions are appropriate (for example, which communication links to secure, databases requiring protection).
2. **Security Administration**—Vulnerabilities emerge because many systems lack a properly structured security policy (security administration is notoriously lax in the case of control systems), equipment and system implementation guides, configuration management, training, and enforcement and compliance auditing.
3. **Architecture**—many common practices negatively affect SCADA security. For example, while it is convenient to use SCADA capabilities for other purposes such as fire and security systems, these practices create single points of failure. Also, the connection of SCADA networks to other automation systems and business networks introduces multiple entry points for potential adversaries.
4. **Network** (including communication links)—legacy systems' hardware and software have very limited security capabilities, and the vulnerabilities of contemporary systems (based on modern information technology) are publicized. Wireless and shared links are susceptible to eavesdropping and data manipulation.
5. **Platforms**—many platform vulnerabilities exist, including retention of default configurations, poor password practices, shared accounts, inadequate protection for hardware, and nonexistent security monitoring controls. In most cases, important security patches are not installed, often due to concern about negatively impacting system operation; in some cases technicians are contractually forbidden from updating systems by their vendor agreements.

The following incident helps to illustrate some of the risks associated with SCADA vulnerabilities.
- During the course of conduction a vulnerability assessment, a contractor stated that personnel from his company penetrated the information system of a utility within minutes. Contractor personnel drove to a remote substation and noticed a wireless network antenna. Without leaving their vehicle, they plugged in their wireless radios and connected to the network within 5 minutes. Within 20 minutes they had mapped the network, including SCADA equipment, and accessed the business network and data. This illustrates that, as a cyber security advisor from Sandia National Laboratories specializing in SCADA stated, utilities are moving to wireless communication without understanding the added risks.

The Increasing Risk

According to GAO (2015), historically, security concerns about control systems (SCADA included) were related primarily to protecting against physical attack and misuse of refining and processing sites or distribution and holding facilities. However, more recently there has been a growing recognition that control systems are now vulnerable to cyber attacks from numerous sources, including hotel governments, terrorist groups, disgruntled employee, and other malicious intruders.

In addition to the control system vulnerabilities mentioned earlier, several factors have contributed to the escalation of risk to control systems, including (1) the adoption of standardized technologies with known vulnerabilities, (2) the connectivity of control systems to other networks, (3) constraints on the implementation of existing security technologies and practices, (4) insecure remote connections, and (5) the widespread availability of technical information about control systems.

Adoption of Technology with Known Vulnerabilities

When a technology is not well known, not widely used, not understood or publicized, it is difficult to penetrate it and thus disable it. Historically, proprietary hardware, software, and network protocols made it difficult to understand how control systems operated—and therefore how to hack into them. Today, however, to reduce costs and improve performance, organizations have been transitioning from proprietary systems to less-expensive, standardized technologies such as Microsoft Windows and Unix-like operating systems and the common networking protocols used by the Internet. These widely used standardized technologies have commonly known vulnerabilities, and sophisticated and effective exploitation tools are widely available and relatively easy to use. As a consequence, both the number of people with the knowledge to wage attacks and the number of systems subject to attack have increased. Also, common communication protocols and the emerging use of Extensible Markup Language (commonly referred to as XML) can make it easier for a hacker to interpret the content of communications among the components of a control system.

Control systems are often connected to other networks—enterprises often integrate their control system with their enterprise networks. This increased connectivity has significant advantages, including providing decision makers with access to real-time information and allowing engineers to monitor and control the process control system from different points on the enterprise network. In addition, the enterprise networks are often connected to the networks of strategic partners and to the Internet. Further, control systems are increasingly using wide area networks and the Internet to transmit data to their remote or local stations and individual devices. This convergence of control networks with public and enterprise networks potentially exposes the control systems to additional security vulnerabilities. Unless appropriate security controls are deployed in the enterprise network and the control system network, breaches in enterprise security can affect the operation of controls system.

According to industry experts, existing security technologies, as well as strong user authentication and patch management practices, are generally not implemented in control systems because control systems operate in real time, typically are not designed with cyber security in mind, and usually have limited processing capabilities.

Existing security technologies such as authorization, authentication, encryption, intrusion detection, and filtering of network traffic and communications require more bandwidth, processing power, and memory than control system components typically have. Because controller stations are generally designed to do specific tasks, they use low-cost, resource-constrained microprocessors. In fact, some devices in the electrical industry still use the Intel 8088 processor, introduced in 1978. Consequently, it is difficult to install existing security technologies without seriously degrading the performance of the control system.

Further, complex passwords and other strong password practices are not always used to prevent unauthorized access to control systems, in part because this could hinder a rapid response to safety procedures during an emergency. As a result, according to experts, weak passwords that are easy to guess, shared, and infrequently changed, including the use of default passwords or even no password at all, are reportedly common in control systems.

In addition, although modern control systems are based on standard operating systems, they are typically customized to support control system applications. Consequently, vendor-provided software patches are generally either incompatible or cannot be implemented without compromising service shutting down "always-on" systems or affecting interdependent operations.

Potential vulnerabilities in control systems are exacerbated by insecure connections. Organizations often leave access links—such as dial-up modems to equipment and control information—open for remote diagnostics, maintenance, and examination of system status. Such links may not be protected with authentication of encryption, which increases the risk that hackers could us these insecure connections to break into remotely controlled systems. Also, control systems often use wireless communications systems, which are especially vulnerable to attack, or leased lines that pass through commercial telecommunications facilities. Without encryption to protect data as it flows through these insecure connections or authentication mechanisms to limit access, there is limited protection for the integrity of the information being transmitted.

Public information about infrastructures and control systems is available to potential hackers and intruders. The availability of this infrastructure and vulnerability data was demonstrated by a university graduate student, whose dissertation reportedly mapped every business and industrial sector in the American economy to the fiber optic network that connects them—using material that was available publicly on the Internet, none of which was classified.

Cyber Threats to Control Systems

There is a general consensus—and increasing concern—among government officials and experts on control systems about potential cyber threats to the control systems that govern our critical infrastructures. As components of control systems increasingly make critical decisions that were once made by humans, the potential effect of a cyber threat becomes more devastating. Such cyber threats could come from numerous sources, ranging from hostile governments and terrorist groups to disgruntled employees and other malicious intruders.

In July 2002, National Infrastructure Protection Center (NIPC) reported that the potential for compound cyber and physical attacks, referred to as "swarming attacks," is an emerging threat to the U.S. critical infrastructure. As NIPC reports, the effects of a swarming attack include slowing or complicating the response to a physical attack. For instance, a cyber attack that disabled the water supply or the electrical system in conjunction with a physical attack could deny emergency services the necessary resources to manage the consequences—such as controlling fires, coordinating actions, and generating light.

Control systems, such as SCADA, can be vulnerable to cyber attacks. Entities or individuals with malicious intent might take one or more of the following actions to successfully attack control systems:

- Disrupt the operation of control systems by delaying or blocking the flow of information through control networks, thereby denying availability of the networks to control system operations
- Make unauthorized changes to programmed instructions in PLCs, RTUs, or DCS controllers, change alarm thresholds, or issue unauthorized commends to control equipment, which could potentially result in damage to equipment (if tolerances are exceeded), premature shutdown of processes (such as prematurely shutting down transmission lines), or even disabling of control equipment
- Send false information to control system operators either to disguise unauthorized changes or to initiate inappropriate actions by system operators
- Modify the control system software, producing unpredictable results
- Interfere with the operation of safety systems

In addition, in control systems that cover a wide geographic area, the remote sites are often unstaffed and may not be physically monitored. If such remote systems are physically breached, the attackers could establish a cyber connection to the control network.

SECURING CONTROL SYSTEMS

Several challenges must be addressed to effectively secure control systems against cyber threats. These challenges include: (1) the limitations of current security technologies in securing control systems; (2) the perception that securing control systems may not be economically justifiable; and (3) the conflicting priorities within organizations regarding the security of control systems.

A significant challenge in effectively securing control systems is the lack of specialized security technologies for these systems. The computing resources in control systems that are needed to perform security functions tend to be quite limited, making it very difficult to use security technologies within control system networks without severely hindering performance.

Securing control systems may not be perceived as economically justifiable. Experts and industry representatives have indicated that organizations may be reluctant to spend more money to secure control systems. Hardening the security of control systems would require industries to expend more resources, including acquiring more personnel, providing training for personnel, and potentially prematurely replacing current systems that typically have a lifespan of about 20 years.

Finally, several experts and industry representatives indicated that the responsibility for securing control systems typically includes two separate groups: IT security personnel and control system engineers and operators. IT security personnel tend to focus on securing enterprise systems, while control system engineers and operators tend to be more concerned with the reliable performance of their control systems. Further, they indicate that, as a result, those two groups do not always fully understand each other's requirements and collaborate to implement secure control systems.

STEPS TO IMPROVE SCADA SECURITY

The President's Critical Infrastructure Protection Board and the Department of Energy (DOE) have developed the steps outlined below to help organizations improve the security of their SCADA networks. DOE (2001) points out that these steps are not meant to be prescriptive or all-inclusive. However, they do address essential actions to be taken to improve the protection of SCADA networks. The steps are divided into two categories: specific actions to improve implementation, and actions to establish essential underlying management processes and policies.

21 STEPS TO INCREASE SCADA SECURITY (DOE, 2001)

The following steps focus on specific actions to be taken to increase the security of SCADA networks:

1. **Identify all connections to SCADA networks.**
   Conduct a thorough risk analysis to assess the risk and necessity of each connection to the SCADA network. Develop a comprehensive understanding of all connections to the SCADA network, and how well those connections are protected. Identify and evaluate the following types of connections:

- Internal local area and wide area networks, including business networks
- The Internet
- Wireless network devices, including satellite uplinks
- Modem or dial-up connections
- Connections to business partners, vendors, or regulatory agencies

2. **Disconnect unnecessary connections to the SCADA network.**

   To ensure the highest degree of security of SCADA systems, isolate the SCADA network from other network connections to as great a degree as possible. Any connection to another network introduces security risks, particularly if the connection creates a pathway from or to the Internet. Although direct connections with other networks may allow important information to be passed efficiently and conveniently, insecure connections are simply not worth the risk; isolation of the SCADA network must be a primary goal to provide needed protection. Strategies such as utilization of "demilitarized zones" (DMZs) and data warehousing can facilitate the secure transfer of data from the SCADA network to business networks. However, they must be designed and implemented properly to avoid introduction of additional risk through improper configuration.

3. **Evaluate and strengthen the security of any remaining connections to the SCADA networks.**

   Conduct penetration testing or vulnerability analysis of any remaining connections to the SCADA network to evaluate the protection posture associated with these pathways. Use this information in conjunction with risk management processes to develop a robust protection strategy for any pathways to the SCADA network. Since the SCADA network is only as secure as its weakest connecting point, it is essential to implement firewalls, intrusion detection systems (IDSs), and other appropriate security measures at each point of entry. Configure firewall rules to prohibit access from and to the SCADA network, and be as specific as possible when permitting approved connections. For example, an independent system operator (ISO) should not be granted "blanket" network access simply because there is a need for a connection to certain components of the SCADA system. Strategically place IDSs at each entry point to alert security personnel of potential breaches of network security. Organization management must understand and accept responsibility or risks associated with any connection to the SCADA network.

4. **Harden SCADA networks by removing or disabling unnecessary services.**

   SCADA control servers built on commercial or open-source operating systems can be exposed to attack default network services. To the greatest degree possible, remove or disable unused services and network demons to reduce the risk of direct attack. This is particularly important when SCADA networks are interconnected with other networks. Do not permit a service or feature on a SCADA network unless a thorough risk assessment of the consequences of allowing the service/feature shows that the benefits of the service/feature far outweigh the potential for vulnerability exploitation.

Examples of services to remove from SCADA networks include automated meter reading/remote billing systems, email services, and Internet access. An example of a feature to disable is remote maintenance. Additionally, work closely with SCADA vendors to identify secure configurations and coordinate any and all changes to operational systems to ensure that removing or disabling services does not cause downtime, interruption of service, or loss of support.

5. **Do not rely on proprietary protocols to protect your system.**

    Some SCADA systems are unique, proprietary protocols for communications between field devices and servers. Often the security of SCADA systems is based solely on the secrecy of these protocols. Unfortunately, obscure protocols provide very little "real" security. Do not rely on proprietary protocols or factor default configuration setting to protect your system. Additionally, demand that vendors disclose any backdoors or vendor interfaces to your SCADA systems, and expect them to provide systems that are capable of being secured.

6. **Implement the security features provided by device and system vendors.**

    Older SCADA systems (most systems in use) have no security features whatsoever; SCADA system owners must insist that their system vendor implement security features in the form of product patches or upgrades. Some newer SCADA devices are shipped with basic security features, but these are usually disabled to ensure ease of installation.

    Analyze each SCADA device to determine whether security features are present. Additionally, factory default security settings (such as in computer network firewalls) are often set to provide maximum usability, but minimal security. Set all security features to provide the maximum security only after a thorough risk assessment of the consequences of reducing the security level.

7. **Establish strong controls over any medium that is used as a backdoor into the SCADA network.**

    Where backdoors or vendor connections do exist in SCADA systems, strong authentication must be implemented to ensure secure communications. Modems, wireless, and wired networks used for communications and maintenance represent a significant vulnerability to the SCADA network and remote sites. Successful "war dialing" or "war driving" attacks could allow an attacker to bypass all of other controls and have direct access to the SCADA network or resources. To minimize the risk of such attacks, disable inbound access and replace it with some type of callback system.

8. **Implement internal and external intrusion detection systems and establish 24-hours-a-day incident monitoring.**

    To be able to effectively respond to cyber attacks, establish an intrusion detection strategy that includes alerting network administrators of malicious network activity originating from internal or external sources. Intrusion detection system monitoring is essential 24 hours a day; this capability can be easily set up through a pager. Additionally, incident response procedures must be in place to allow an effective response to any attack. To complement network monitoring, enable logging on all systems and audit system logs daily to detect suspicious activity as soon as possible.

9. **Perform technical audits of SCADA devices and networks, and any other connected networks, to identify security concerns.**

    Technical audits of SCADA devices and networks are critical to ongoing security effectiveness. Many commercial and open-sourced security tools are available that allow system administrators to conduct audits of their systems/networks to identify active services, patch level, and common vulnerabilities. The use of these tools will not solve systemic problems, but will eliminate the "paths of least resistance" that an attacker could exploit. Analyze identified vulnerabilities to determine their significance, and take corrective actions as appropriate. Track corrective actions and analyze this information to identify trends. Additionally, retest systems after corrective actions have been taken to ensure that vulnerabilities were actually eliminated. Scan non-production environments actively to identify and address potential problems.

10. **Conduct physical security surveys and assess all remote sites connected to the SCADA network to evaluate their security.**

    Any location that has a connection to the SCADA network is a target, especially unmanned or unguarded remote sites. Conduct a physical security survey and inventory access points at each facility that has a connection to the SCADA system. Identify and assess any source of information, including remote telephone/computer network/fiber optic cables that could be tapped; radio and microwave links that are exploitable; computer terminals that could be accessed; and wireless local area network access points. Identify and eliminate single points of failure. The security of the site must be adequate to detect or prevent unauthorized access. Do not allow "live" network access points at remote, unguarded sites simply for convenience.

11. **Establish SCADA "Red Teams" to identify and evaluate possible attack scenarios.**

    Establish a "Red Team" to identify potential attack scenarios and evaluate potential system vulnerabilities. Use a variety of people who can provide insight into weaknesses of the overall network, SCADA system, physical systems, and security controls. People who work on the system every day have great insight into the vulnerabilities of your SCADA network and should be consulted when identifying potential attack scenarios and possible consequences. Also, ensure that the risk from a malicious insider is fully evaluated, given that this represents one of the greatest threats to an organization. Feed information resulting from the "Red Team" evaluation into risk management processes to assess the information and establish appropriate protection strategies.

    The following steps focus on management actions to establish an effective cyber security program:

12. **Clearly define cyber security roles, responsibilities, and authorities for managers, system administrators, and users.**

Organization personnel need to understand the specific expectations associated with protecting information technology resources through the definition of clear and logical roles and responsibilities. In addition, key personnel need to be given sufficient authority to carry out their assigned responsibilities. Too often, good cyber security is left to the initiative of the individual, which usually leads to inconsistent implementations and ineffective security. Establish a cyber security organizational structure that defines roles and responsibilities and clearly identifies how cyber security issues are escalated and who is notified in an emergency.

13. **Document network architecture and identify systems that serve critical functions or contain sensitive information that require additional levels of protection.**

    Develop and document robust information security architecture as part of a process to establish an effective protection strategy. It is essential that organizations design their network with security in mind and continue to have a strong understanding of their network architecture throughout its lifecycle. Of particular importance, an in-depth understanding of the functions that the systems perform and the sensitivity of the stored information is required. Without this understanding, risk cannot be properly assessed and protection strategies may not be sufficient. Documenting the information security architecture and its components is critical to understanding the overall protection strategy, and identifying single points of failure.

14. **Establish a rigorous, ongoing risk management process.**

    A thorough understanding of the risks to network computing resources from denial-of-service attacks and the vulnerability of sensitive information to compromise is essential to an effective cyber security program. Risk assessments form the technical basis of this understanding and are critical to formulating effective strategies to mitigate vulnerabilities and preserve the integrity of computing resources. Initially, perform a baseline risk analysis based on current threat assessment to use for developing a network protection strategy. Due to rapidly changing technology and the emergence of new threats on a daily basis, an ongoing risk assessment process is also needed so that routine changes can be made to the protection strategy to ensure it remains effective. Fundamental to risk management is identification of residual risk with a network protection strategy in place and acceptance of that risk by management.

15. **Establish a network protection strategy based on the principle of defense-in-depth.**

    A fundamental principle that must be part of any network protection strategy is defense-in-depth. Defense-in-depth must be considered early in the design phase of the development process, and must be an integral consideration in all technical decision making associated with the network. Utilize technical and administrative controls to mitigate threats from identified risks to as great a degree as possible at all levels of the network. Single points of failure must be avoided, and cyber security defense must be layered to limit and contain the impact of any security incidents. Additionally, each layer must be protected against other systems at the same layer. For example, to protect against the inside threat, restrict users to access only those resources necessary to perform their job functions.

16. **Clearly identify cyber security requirements.**

    Organizations and companies need structured security programs with mandated requirements to establish expectations and allow personnel to be held accountable. Formalized policies and procedures are typically used to establish and institutionalize a cyber security program. A formal program is essential to establish a consistent, standards-based approach to cyber security through an organization and eliminates sole dependence on individual initiative. Policies and procedures also inform employees of their specific cyber security responsibilities and the consequences of failing to meet those responsibilities. They also provide guidance regarding actions to be taken during a cyber security incident and promote efficient and effective actions during a time of crisis. As part of identifying cyber security requirements, include user agreements and notification and warning banners. Establish requirements to minimize the threat from malicious insiders, including the need for conducting background checks and limiting network privileges to those absolutely necessary.

17. **Establish effective configuration management processes.**

    A fundamental management process needed to maintain a secure network is configuration management. Configuration management needs to cover both hardware configurations and software configurations. Changes to hardware or software can easily introduce vulnerabilities that undermine network security. Processes are required to evaluate and control any change to ensure that the network remains secure. Configuration management begins with well-tested and documented security baselines for your various systems.

18. **Conduct routine self-assessments.**

    Robust performance evaluation processes are needed to provide organizations with feedback on the effectiveness of cyber security policy and technical implementation. A sign of a mature organization is that it is able to self-identify issues, conduct root cause analyses, and implement effective corrective actions that address individual and systemic problems. Self-assessment processes that are normally part of an effective cyber security program include routine scanning for vulnerabilities, automated auditing of the network, and self-assessments of organizational and individual performance.

19. **Establish system backups and disaster recovery plans.**

    Establish a disaster recovery plan that allows for rapid recovery from any emergency (including a cyber attack). System backups are an essential part of any plan and allow rapid reconstruction of the network. Routinely exercise disaster recovery plans to ensure that they work and that personnel are familiar with them. Make appropriate changes to disaster recovery plans based on lessons learned from exercises.

20. **Senior organizational leadership should establish expectations for cyber security performance and hold individuals accountable for their performance.**

    Effective cyber security performance requires commitment and leadership from senior managers in the organization. It is essential that senior management establish an expectation for strong cyber security and communicate this to their subordinate managers throughout the organization. It is also essential that senior organizational leadership establish a structure for implementation of a cyber security program. This structure will promote consistent implementation and the ability to sustain a strong cyber security program. It is then important for individuals, including managers, system administrators, technicians, and users/operators, to be held accountable for their performance as it relates to cyber security.

21. **Establish policies and conduct training to minimize the likelihood that organizational personnel will inadvertently disclose sensitive information regarding SCADA system design, operations, or security controls.**

    Release data related to the SCADA network only on a strict, need-to-know basis, and only to persons explicitly authorized to receive such information. "Social engineering," the gathering of information about a computer or computer network via questions to naïve users, is often the first step in a malicious attack on computer networks. The more the information revealed about a computer or computer network, the more vulnerable the computer/network is. Never divulge data revealed to a SCADA network, including the names and contact information about the system operators/administrators, computer operating systems, and/or physical and logical locations of computers and network systems over telephones or to personnel unless they are explicitly authorized to receive such information. Any requests for information by unknown persons need to be sent to a central network security location for verification and fulfillment. People can be a weak link in an otherwise secure network. Conduct training and information awareness campaigns to ensure that personnel remain diligent in guarding sensitive network information, particularly their passwords.

REFERENCES AND RECOMMENDED READING

Associated Press (AP). 2009. Goal: Disrupt. *The Virginian-Pilot*, Norfolk, VA, 04 April 2009.

Brown, A.S. 2008. *SCADA vs. the Hackers*. American Society of Mechanical Engineers. Accessed 05 October 2019 @ http://memagazine.org/backissues/dec02/features/scadavs/.

DOE. 2001. *21 Steps to Improve Cyber Security of SCADA Networks*. Washington, DC: Department of Energy.

Ezell, B.C. 1998. *Risks of Cyber Attack to Supervisory Control and Data Acquisition*. Charlottesville, VA: University of Virginia.

FEMA. 2008. *FEMA452: Risk Assessment: A How to Guide*. Accessed 05/01/08 @fema.gov/library/file?type=published/filetofile.

FEMA. 2015. *Protecting Critical Infrastructure Against Insider Threats*. Accessed 17 April 2019 @ http://emilms.fema.gov/IS0915/IABsummary.htm.

FBI. 2000. *Threat to Critical Infrastructure*. Washington, DC: Federal Bureau of Investigation.
FBI. 2007. *Ninth Annual Computer Crime and Security Survey*. FBI: Computer Crime Institute and Federal Bureau of Investigations, Washington, DC.
FBI. 2014. *Protecting Critical Infrastructure and the Importance of Partnerships*. Accessed 19 December 2019 @ https://fib.gov/news/speeches/protecitng-critical-infrastruce-and t-the-importantce-o.
GAO. 2003. *Critical Infrastructure Protection: Challenges in Securing Control System*. Washington, DC: United States General Accounting Office.
GAO. 2015. *Critical Infrastructure Protection: Sector-Specific Agencies Need to Better Measure Cybersecurity Progress*. Washington, DC: United States Government Accountability Office.
Gellman, B. 2002. Cyber-Attacks by Al Qaeda Feared: Terrorists at Threshold of Using Internet as Tool of Bloodshed, Experts Say. *Washington Post*, June 27, p. A01.
Minter, J.G. 1996. Prevention Chemical Accidents Still a Challenge. *Occupational Hazards*, September, xix, 22–30.
National Infrastructure Advisory Council. 2008. *First Report and Recommendations on the Insider Threat to Critical Infrastructure*. Department of Homeland Security, Washington, DC.
NIPC. 2002. *National Infrastructure Protection Center Report*. Washington, DC: National Infrastructure Protection Center.
Spellman, F.R. 1997. *A Guide to Compliance for PSM/RMP*. Lancaster, PA: Technomic Publishing Company.
Stamp, J., et al. 2003. *Common Vulnerabilities in Critical Infrastructure Control Systems*. 2nd ed. Sandia National Laboratories.
U.S. Department of Energy. 2010. *Energy Sector-Specific Plan: An Annex to the National Infrastructure Protection Plan*. Washington, DC: USDOE.
US DOE. 2002. *Vulnerability Assessment Methodology: Electric Power Infrastructure*. Department of Energy, Washington, DC.
USEPA. 2005. EPA Needs to Determine What Barriers Prevent Water Systems from Securing Known SCADA Vulnerabilities. In: Harris, J. (ed.), *Final Briefing Report*. Washington, DC: USEPA.
Warwalking. 2003. Warwalking, Accessed 05 September 2019 @ http://warwalking.tribe.net.
Wiles, J., et al. 2007. *Techno Security's™ Guide to Securing SCADA*. Burlington, MA: Elsevier, Inc.
Young, M.A. 2004. *SCADA Systems Security*. SANS Institute, Sans Institute Rockville, Maryland.

10 IT Security Action Plan

Never underestimate the time, expense, blood, sweat, and effort a terrorist will apply to compromise the security of any industrial facility.

Note: In this book we use the terms information technology (IT), cyber space, and digital network interchangeably.

IT SECURITY ACTION ITEMS

Information Technology security policies should follow good design and governance practices—not so long that they become unusable, not so vague that they become meaningless—and should be reviewed on a regular basis to ensure that they stay pertinent as needs change.

All organizations should develop and maintain clear and robust polices for safeguarding critical business data and sensitive information, protecting their reputation and discouraging inappropriate behavior by employees. These robust policies for safeguarding digital sensitive information should include (based on FCC, 2017):

- Policy development and management
- Scams and fraud
- Network security
- Website security
- Email
- Mobile devices
- Employees
- Facility security
- Operational security
- Payment cards
- Incident response and reporting

Note: Implementing the following recommended actions to protect your IT sector systems is highly advised. However, if organization funds are available to hire reputable professional IT security professionals to tailor security needs to your system this is always the best course of action; moreover, many of the items detailed in the following should or would be included and accomplished in large companies by outside and/or inside computer experts.

Policy Development and Management

Establish Security Roles and Responsibilities

One of the least expensive and most effective means of preventing serious IT security incidents is to establish a policy that clearly defines the separation of roles and responsibilities with regard to systems and the information they contain. Such polices need to clearly identify company data ownership and employee roles for security oversight and their inherent privileges, including:

- Necessary roles, and the privileges and constraints accorded to those roles
- The types of employees who should be allowed to assume the various roles
- Mandatory review of how long an employee may hold a role before access rights
- Adoption of a specific role as dictated by the circumstances should employees hold multiple roles

Depending on the types of data regularly handled by your business, it may also make sense to create separate policies governing who is responsible for certain types of data.

Establish an Employee Internet Usage Policy

The limits on employee Internet usage in the workplace vary widely from business to business. Your guidelines should allow employees the maximum degree of freedom they require to be productive (short breaks to surf the web or perform personal tasks online have been shown to increase productivity). However, at the same time, employees must keep in mind that rules of behavior are necessary to ensure that they are aware of boundaries, both to keep them safe and to keep the company successful. Consider, for example:

- Personal breaks to surf the web should be limited to a reasonable amount of time and to certain types of activities.
- If a web filtering system is used, employees should have clear knowledge of how and why their web activities will be monitored, and what types of sites are deemed unacceptable by your policy.
- Workplace rules of behavior should be clear, concise, and easy to follow. Employees should feel comfortable performing both personal and professional tasks online without making judgment calls as to what may or may not be deemed appropriate. Businesses may want to include a splash warning upon network sign-on that advises the employees of the businesses' Internet usage polices so that all employees are on notice.

Establish a Social Media Policy

Using technical or procedural solutions to control access to social networking applications introduces a number of risks that are difficult to address. A strong social media policy is crucial for any business that seeks to use social networking to

IT Security Action Plan

promote its activities and communicate with its customers. At a minimum, a social media policy should clearly include the following:

- Specific guidance on when to disclose company activities using social media, and what kinds of details can be discussed in a public forum
- Additional rules of behavior for employees using personal social networking accounts to make clear what kinds of discussion topics or posts could cause risk for the company
- Guidance on the acceptability of using a company email address to register for, or get notices from, social media sites
- Guidance on selecting long and strong passwords for social networking accounts, since very few social media sites enforce strong authentication policies for users

Last but not least, all users of social media need to be aware of the risks associated with social networking tools and the types of data that can be automatically discovered online when using social media. Taking the time to educate your employees on the potential pitfalls of social media use, especially in tandem with geo-location services (i.e., the latitude and longitude of a particular location), may be the most beneficial social networking security practice of all.

Identify Potential Reputation Risks

Potential risks to a company's reputation are real and therefore a strategy should be developed to mitigate those risks via polices or other measures. Specific types of reputation risks include:

- Being impersonated online by a criminal organization (e.g., an illegitimate website spoofing a business name and copying site design, then attempting to defraud potential customers via phishing scams or other methods)
- Having sensitive company or customer information leaked to the public via the web
- Having sensitive or inappropriate employee actions made public via the web or social media sites

All businesses should set a policy for managing these types of risks and plans to address such incidents if and when they occur. Such a policy should cover a regular process for identifying potential risks to the company's reputation in cyberspace, practical measures to prevent those risks from materializing, and reference plans to respond and recover from potential incidents as soon as they occur.

SCAMS AND FRAUD

New telecommunications technologies may offer countless opportunities for business, but they also offer cyber criminals many new ways to victimize businesses, scam customers, and hurt reputations.

Train Employees to Recognize Social Engineering

Social engineering, also known as "pretexting," is used by many criminals, both online and offline, to trick unsuspecting people into giving away their personal information and/or installing malicious software onto their computers, devices, or networks. Social engineering is successful because the bad guys are doing their best to make their work look and sound legitimate, sometimes even helpful, which makes it easier to deceive users.

Most offline social engineering occurs over the telephone, but it frequently occurs online as well. Information gathered from social networks or posted on websites can be enough to create a convincing ruse to trick employees. For example, LinkedIn profiles, Facebook posts, and twitter messages can allow a criminal to assemble detailed dossiers on employees. Teaching people the risks involved in sharing personal or business details on the Internet can help partner with staff to prevent both personal and organizational losses.

Many criminals use social engineering tactics to get individuals to voluntarily install malicious computer software such as fake antivirus, thinking they are doing something that will help make them more secure. Users who are tricked into loading malicious programs on their computers may be providing remote control capabilities to an attacker, unwittingly installing software that can steal financial information or simply try to sell them fake security software.

Protect against Online Fraud

Online fraud takes on many guises that can impact everyone, including businesses and their employees. It is helpful to maintain consistent and predictable online messaging when communicating with customers to prevent others from impersonating the company.

Be sure to never request personal information or account details through email, social networking, or other online messaging platforms. Let customers know their personal information will never be requested through such channels and instruct them to contact the person or company directly should they have any concerns.

Protect against Phishing

Phishing is the technique used by online criminals to trick people into thinking they are dealing with a trusted website or other entity. Businesses face this threat from two directions—phishers may be impersonating them to take advantage of unsuspecting customers, and phishers may be trying to steal their employees' online credentials.

Businesses should ensure that their online communications never ask their customers to submit sensitive information via email. Businesses should make clear in their communications that the business will never ask for personal information via email so that if someone targets customers, they may realize it is a scam.

The best defense against users being tricked into handing over their usernames and passwords to cyber criminals is employee awareness. Employees need to be informed that they should never respond to incoming messages requesting private information. Also, to avoid being led to a fake site, they should know to never click

IT Security Action Plan 157

on a link sent by email from an untrustworthy source. Employees who need to access a website link sent from a questionable source should open an Internet browser window and manually type in the site's web address to make sure the emailed link is not maliciously redirecting to a dangerous site. Keep in mind that this is especially critical for protecting online banking accounts belonging to your organization. Criminals target business banking accounts.

Don't Fall for Fake Antivirus Offers

"Scareware" (fake antivirus) and other rogue online security scams have been behind some of the most successful online frauds in recent times. Make sure your organization has a policy in place explaining what the procedure is if an employee's computer becomes infected by a virus.

Employees need to be trained to recognize a legitimate warning message and to properly notify the IT team if something bad or questionable has happened. If possible, business computers should be configured to not allow regular users to have administrative access. This will minimize the risk of them installing malicious software and condition users that adding unauthorized software to work computers is against policy.

Protect against Malware

Businesses can experience a compromise through the introduction of malicious software, or malware, that tracks a user's keyboard strokes, also known as *key logging*.

Many organizations are falling victim to key-logging malware being installed on computer systems in their environment. Once installed, the malware can record keystrokes made on a computer, allowing bad guys to see passwords, credit card numbers, and other confidential data. Keeping security software up to date and patching computers regularly will make it more difficult for this type of malware to infiltrate business networks.

Develop a Layered Approach to Guard against Malicious Software

Despite progress in creating more awareness of security threats on the Internet, malware authors are not giving up. Based on research, it is found that more than 100,000 malicious software samples are seen every day (FCC, 2017).

Effective protection against viruses, Trojans, and other malicious software requires a layered approach to company defenses. The company must install antivirus software but should also deploy a combination of multiple techniques to keep the workplace environment safe. Moreover, the use of thumb drives and other removable media must be guarded against security threats; as they may contain pre-installed malicious software that can infect the organization's computer(s), ensure that the source of the removable media devices is trustworthy before installing them.

Whenever a combination of web filtering, antivirus signature protection, proactive malware protection, firewalls, strong security policies, and employee training are used, the risk of infection is significantly reduced. Keeping protection software up to date along with your operating system and applications increases the safety of your systems.

Verify the Identity of Telephone Information Seekers

The telephone is the preferred communication device used by social engineering criminals. Information gathered through social networks and information posted on websites can be enough to create a convincing ruse to trick company employees. Ensure employees are trained to never disclose customer information, usernames, passwords, or other sensitive details to incoming callers. When someone requests information, always contact the person back using a known phone number or email account to verify the identity and validity of the individual and their request.

NETWORK SECURITY

Securing an organization's network consists of (1) identifying all devices and connections on the network; (2) setting boundaries between the company's systems and others, and (3) enforcing controls to ensure that unauthorized access, misuse, or denial-of-service events can be thwarted or rapidly contained and recovered from if they do occur.

Secure Internal Network and Cloud Services

A company's network should be separated from the public Internet by strong user authentication mechanism and policy enforcement systems such as firewalls and web filtering proxies. Additional monitoring and security solutions, such as antivirus software and intrusion detection systems, should also be employed to identify and stop malicious code or unauthorized access attempts.

Internal Network

After identifying the boundary points on the company's network, each boundary should be evaluated to determine what types of security controls are necessary and how they can be best deployed. Border routers should be configured to only route traffic to and from a company's public IP addresses, firewalls would be deployed to restrict traffic only to and from the minimum set of necessary services, and intrusion prevention systems should be confirmed to monitor for suspicious activity crossing the network perimeter. In order to prevent bottlenecks, all security systems you deploy to your company's network perimeter should be capable of handling the bandwidth that your carrier provides.

Cloud-Based Services

Cloud computing is a model for enabling convenient, on-demand network access to a shared pool of configurable computing resources. The terms of service with all cloud service providers should be consulted to ensure that the company's information and activities are protected with the same degree of security that the company would intend to provide on its own. Businesses should request security and auditing from their cloud service providers as applicable to the business's needs and concerns. Service-level agreements of system restoration and reconstitution time must be reviewed and understood. Businesses should also inquire about additional services a cloud service can provide. These services may include backup-and-restore services and encryption services, which may be very attractive to businesses.

Develop Strong Password Policies

Generally speaking, two-factor authentication methods, which require two types of evidence that the employee (or anyone else) is who he or she claims to be, are safer than using static passwords for authentication. One common example is a personal security token that displays changing passcodes to be used in conjunction with an established password. However, two-factor systems may not always be possible or practical for your company.

Company password policies should encourage employees to use the strongest passwords possible without creating the need or temptation to reuse passwords or write them down. That means passwords should be random, complex, and long (at least 10 characters); changed regularly; and closely guarded by those who know them.

Secure and Encrypt the Organization Wi-Fi

Wireless Access Control

A company may operate a wireless local area network (WLAN) for the use of customers, guests, and visitors. If this is the case, it is important that such a WLAN be kept separate from the main company network so that traffic from the public network cannot traverse the company's internal systems at any point.

Internal, non-public WLAN access should be restricted to specific devices and specific users to the greatest extent possible while meeting your company's business needs. Where the internal WLAN has less stringent access controls than your company's wired network, dual connections—where a device is able to connect to both the wireless and the wired networks simultaneously—should be prohibited by technical controls on each such capable device (e.g., BIOS-level LAN/WLAN switch settings). All users should be given unique credentials with preset expiration dates to use when accessing the internal WLAN.

Wireless Encryption

Due to demonstrable security flaws that are known to exist in older forms of wireless encryption, a company's internal WLAN should only employ Wi-Fi Protected Access 2 (WPA2) encryption.

Encrypt Sensitive Company Data

Encryption should be used to protect any data that a company considers sensitive, in addition to meeting applicable regulatory requirements on information safeguarding. Different encryption schemes are appropriate under different circumstances. However, applications that comply with the OpenPGP standard, such as PGP (Pretty Good Privacy encryption program) and GnuPG, provide a wide range of options for securing data on disk as well as in transit. If you choose to offer security transactions via your company's website, consult with your service provider about available options for a Secure Socket Layer (SSL) certificate for a business site.

Regularly Update All Applications

All systems and software, including networking equipment, should be updated in a timely fashion as patches and firmware upgrades become available. Use automatic updating services whenever possible, especially for security systems such as anti-malware applications, web filtering tools, and intrusion prevention systems.

Set Safe Web Browsing Rules

An organization's internal network should only be able to access those services and resources on the Internet that are essential for the business and the needs of employees. Use the safe browsing features included with modern web browsing software and a web proxy to ensure that malicious or unauthorized sites cannot be accessed from your internal network.

If Remote Access Is Enabled, Make Sure It Is Secure

If an organization needs to provide remote access to its internal network over the Internet, one popular and secure option is to use a secure Virtual Private Network (VPN) system accompanied by strong two-factor authentication, using either hardware or software tokens (FCC, 2017).

WEB SECURITY

No amount of hyperbole or emphasis is needed when we say that website security is more important than ever. Web servers, which host the data and other content available to customers on the Internet, are often the most targeted and attacked components of a company's network. Cyber criminals are constantly looking for improperly secured websites to attack, and many customers say website security is a top consideration when they choose to shop online. As a result, it is essential to secure severs and the network infrastructure that supports them. The consequences of a security breach are great: loss of revenues, damage to credibility, legal liability, and loss of customer trust. The following are examples of specific security threats to web servers (FCC, 2017):

- Cyber criminals may exploit software bugs in the web server, underlying operating system, or active content to gain unauthorized access to the web server. Examples of unauthorized access include gaining access to files or folders that were not meant to be publicly accessible and being able to execute commands and/or install malicious software on the web server.
- Denial-of-service attacks may be directed at the web server or its supporting network infrastructure to prevent or hinder a company's website users from making use of its services.
- Sensitive information on the web server may be read or modified without authorization.
- Sensitive information on backend databases that are used to support interactive elements of a web application may be compromised through the injection of unauthorized software commands.
- Sensitive unencrypted information transmitted between the web server and the browser may be intercepted.
- Information on the web server may be changed for malicious purposes. Website defacement is a commonly reported example of this threat.

IT Security Action Plan

- Cyber criminals may gain unauthorized access to resources elsewhere in the organization's network via a successful attack on the web server.
- Cyber criminals may also attack external entities after compromising a web server. The attacks can be launched directly (e.g., from the compromised server against an external server) or indirectly (e.g., placing malicious content on the compromised web server that attempts to exploit vulnerability in the web browsers of users visiting the site).
- The server may be used as a distribution point for attack tools, pornography, or illegally copied software.

Carefully Plan and Address the Security Aspects of the Deployment of a Public Web Server

Because it is much more difficult to address security once deployment and implementation have occurred, security should be considered from the initial planning stage. Organizations are more likely to make decisions about configuring computers appropriately and consistently when they develop and use a detailed, well-designed deployment plan. Developing such a plan will support web server administrators in making the inevitable tradeoff decisions between usability, performance, and risk.

Businesses also need to consider the human resource requirements for the deployment and continued operation of the web server and supporting infrastructure. The following points in a deployment plan:

- Types of personnel required—for example, system and web server administrators, webmaster, network administrators, and information systems security personnel
- Skills and training required by assigned personnel
- Individual (i.e., the level of effort required of specific personnel types) and collective staffing (i.e., overall level of effort) requirements

Implement Appropriate Security Management Practices and Controls when Maintaining and Operating a Secure Web Server

Appropriate management practices are essential to operating and maintaining a secure web server. Security practices include the identification of the company's information system assets and the development, documentation, and implementation of policies, and guidelines to help ensure the confidentiality, integrity, and availability of information system resources. The following practices and controls are recommended:

- A business-wide information system security policy
- Server configuration and change control and management
- Risk assessment and management
- Standardized software configurations that satisfy the information system security policy
- Security awareness and training

- Contingency, planning, continuity of operations, and disaster recovery planning
- Certification and accreditation

Ensure that Web Server Systems Meet the Organization's Security Requirements

The initial step in securing a web server is securing the underlying operations system. Most commonly available web servers operate on a general-purpose operating system. Many security issues can be avoided if the operating systems underlying web servers are configured appropriately. Default hardware and software configurations are typically set by manufacturers to emphasize features, functions, and ease of use at the expense of security. Because manufacturers are not aware of each organization's security needs, each web server administrator must configure new servers to reflect their business's security requirements and reconfigure them as those requirements change. Using security configuration guides or checklists can assist administrators in security systems consistently and efficiently. Initially securing an operating system initially generally includes the following steps:

- Patch and upgrade the operating system
- Change all default passwords
- Remove or disable unnecessary services and applications
- Configure operating system user authentication
- Configure resource controls
- Install and configure additional security controls
- Perform security testing of the operating system

Ensure the Web Server Application Meets the Organization's Security Requirements

In many respects, the security installation and configuration of the web server application will mirror the operating system process discussed above. The overarching principle is to install the minimal amount of web server services required and eliminate any known vulnerabilities through patches or upgrades. If the installation program installs any unnecessary applications, services, or scripts, they should be removed immediately after the installation process concludes. Securing the web server application generally includes the following steps:

- Patch and upgrade the web server application
- Remove or disable unnecessary services, applications, and sample content
- Configure web server user authentication and access controls
- Configure web server resource controls
- Test the security of the web server application and web content

Ensure that Only Appropriate Content Is Published on the Organization's Website

Typically it is the organization's website that is often one of the first places cyber criminals search for valuable information. In spite of everything, many

IT Security Action Plan

businesses lack a web publishing process or policy that determines what type of information to publish openly, what information to publish with restricted access and what information should not be published to any publicly accessible repository. Some generally accepted examples of what should not be pulled or at least should be carefully examined and reviewed before being published on a public website include:

- Classified or proprietary business information
- Sensitive information relating to business's security
- Medical records
- A business's detailed physical and information security safeguards
- Details about a business's network and information system infrastructure—for example, address ranges, naming conventions, and access numbers
- Information that specifies or implies physical security vulnerabilities
- Detailed plans, maps, diagrams, aerial photographs, and architectural drawings of business buildings, properties, or installations
- Any sensitive information about individuals that might be subject to federal, state, or, in some instances, international privacy laws

Ensure Appropriate Steps Are Taken to Protect Web Content from Unauthorized Access or Modification

Although information available on public websites is intended to be public (assuming a credible review process and policy is in place), it is still important to ensure that information cannot be modified without authorization. Users of such information rely on its integrity even if the information is not confidential. Content on publicly accessible web servers is inherently more vulnerable than information that is inaccessible from the Internet, and this vulnerability means businesses need to protect public web content through appropriate configuration of web server resource controls. Examples of resource control practices include:

- Install or enable only necessary services
- Install web content on a dedicated hard drive or logical partition
- Limit uploads to directories that are not readable by the web server
- Define a single directory for all external scripts or programs executed as part of web content
- Disable the use of hard to symbolic links
- Define a complete web content access matrix identifying which folders and files in the web server document directory are restricted, which are accessible, and by whom
- Disable directory listings
- Deploy user authentication to identify approved users, digital signatures, and other cryptographic mechanisms as appropriate
- Use intrusion detection systems, intrusion prevent systems, and file integrity checkers to spot intrusions and verify web content
- Protect each backed server (i.e., database server or directory server) from command injection attacks

Use Active Content Judiciously after Balancing the Benefits and Risks

Most early websites contained static information, typically in the form of text-based documents. Soon thereafter, interactive elements were introduced to offer new opportunities for user interaction. Unfortunately, these same interactive elements introduced new web-related vulnerabilities. They typically involved dynamically executing code using a large number of inputs, from web page URL parameters to hypertext transfer protocol (HTTP) content and, more recently, extensible markup language (XML) content. Different active content technologies pose different related vulnerabilities, and their risks should be weighed against their benefits. Although most websites use some form of active content generators, they may also deliver some or all of their content in a static form.

Use Authentication and Cryptographic Technologies as Appropriate to Protect Certain Types of Sensitive Data

Public web servers often support technologies for identifying and authenticating users with differing privileges for accessing information. Some of the technologies are based on cryptographic functions that can provide a secure channel between a web browser client and a web server that supports encryption. Web servers may be configured to use different cryptographic algorithms, providing varying levels of security and performance.

Without proper user authentication in place, businesses cannot selectively restrict access to specific information. All information that resides on subject web server is then accessible by anyone with access to the server. In addition, without some process to authenticate the server, users of the public web server will not be able to determine whether the server is the "authentic" web server or a counterfeit version operated by a cyber criminal.

Even with an encrypted channel and an authentication mechanism, it is possible that attackers may attempt to access the site by brute force. Improper authentication techniques can allow attackers to gather valid usernames or potentially gain access to the website. Strong authentication mechanisms can protect against phishing attacks, in which hackers may trick users into providing their personal credentials, and pharming, in which traffic to a legitimate website may be redirected to an illegitimate one. An appropriate level of authentication should be implemented based on the sensitivity of the web server's users and content.

Employ Network Infrastructure to Help Protect Public Web Servers

The network infrastructure (e.g., firewalls, routers, and intrusion detection systems) that supports the web server plays a critical security role. In most configurations, the network infrastructure will be the first line of defense between a public web server and the Internet. Network design alone, though, can't protect a web server. The frequency, sophistication, and variety of web server attacks perpetrated today support the idea that web server security must be implemented through layered and diverse protection mechanisms, an approach sometimes referred to as "defense in depths"

Commit to an Ongoing Process of Maintaining Web Server Security

Maintaining a secure web server requires constant effort, resources, and vigilance. Security administering a web server on a daily basis is essential. Maintaining the security of a web server will usually involve the following steps:

- Configuring, protecting, and analyzing log files
- Backup critical information frequently
- Maintaining a protected authoritative copy of your organization's web content
- Establishing and following procedures for recovering from compromise
- Testing and applying patches in a timely manner
- Testing security periodically

EMAIL

Email has become a critical part of everyday business, from internal management to direct customer support. The benefits associated with email as a primary business tool far outweigh the negatives. However, organizations must be mindful that a successful email platform starts with basic principles of email security to ensure the privacy and protection of customer and business information.

Set up a Spam Email Filter

It has been well documented that spam, phishing attempts, and otherwise unsolicited and unwelcome email often accounts for more than 60 percent of all email that an individual or business receives. Email is the primary method for spreading viruses and malware, and it is one of the easiest to defend against. Consider using email-filtering services that your email service, hosting provider, or other cloud providers offer. A local email-filter application is also an important component of a solid antivirus strategy. Ensure that automatic updates are enabled on your email application, email filter, and antivirus programs. Ensure that filters are reviewed regularly so that important emails and/or domains are not blocked in error.

Train Employees in Responsible Email Usage

The last line of defense for all of your cyber risk efforts lies with the employee, who should use tools such as email and must regulate their responsible and appropriate use for management of information. Technology alone can't make an organization secure. Employees must be trained to identify risks associated with email use, how and when to use email appropriate to their work, and when to seek assistance of professionals. Employee awareness training is available in many forms, including printed media, videos, and online training.

Consider requiring security awareness training for all new employees and refresher courses every year. Simple efforts such as monthly newsletters, urgent bulletins when new viruses are detected, and even posters in common areas to remind employees of key security and privacy to-dos create a work environment that is educated in protecting the business.

Protect Sensitive Information Sent via Email

With its proliferation as a primary tool to communicate internally and externally, business email often includes sensitive information. Whether it is company information that could harm the business or regulated data such as personal health information (PHI) or personally identifiable information (PH), it is important to ensure that such information is only sent and accessed by those who are entitled to see it.

Since email in its native form is not designed to be secure, incidents of mis-addressing or other common accidental forwarding can lead to data leakage. Organizations that handle this type of inflation should consider whether such information should be sent via email, or at least consider using email encryption. Encryption is the process of converting data into unreadable format to prevent disclosure to unauthorized personnel. Only individuals or organizations with access to the encryption key can read the information. Other cloud services offer "Security Web Enabled Drop Boxes" that enable secure data transfer for sensitive information, which is often a better approach to transmitting between companies or customers.

Set a Sensible Email Retention Policy

Another important consideration is the management of email that resides on company messaging systems and your users' computers. From the cost of storage and backup to legal and regulatory requirements, companies should document how they will handle email retention and implement basic controls to help them attain those standards. Many industries have specific rules that dictate how long emails can or should be retained, but the basic rule of thumb is only as long as it supports your business efforts. Many companies implement a 60–90-day retention standard if not compelled by law to another retention period.

To ensure compliance, companies should consider mandatory archiving at a chosen retention cycle end data and automatic permanent email removal after another set point, such as 180–260 days in archives. In addition, organizations should discourage the use of personal folders on employee computers (most often configurable from the email system level), as this will make it more difficult to manage company standards.

Develop an Email Usage Policy

Policies are important for setting expectations with employees or users, and for developing standards to ensure adherence to the company's published policies. Accordingly, the company's policies should be easy to read, understand, define, and enforce. In addition, key areas to address should include what the company email system should and should not be used for, and what data are allowed to be transmitted. Other policy areas should address retention, privacy, and acceptable use.

Depending on the company's business and jurisdiction, there may be a need for email monitoring. The rights of the business and the user should be documented in the policy as well. The policy should be part of the company's general end-user-awareness training and reviewed for updates on a yearly basis.

Mobile Devices

If an organization uses mobile devices to conduct company business, such as accessing company email or sensitive data, pay close attention to mobile security and the potential threats that can expose and compromise a company's business networks. This section describes the mobile threat environment and the practices that businesses can use to help secure devices such as smartphones, tablets, and Wi-Fi enabled laptops.

Many organizations are finding that employees are most productive when using mobile devices, and the benefits are too great to ignore. But while mobility can increase workplace productivity, allowing employees to bring their own mobile devices into the enterprise can create significant security and management challenges.

Data loss and data breaches caused by lost or stolen phones create big challenges, as mobile devices are now used to store confidential business information and access the corporate network. Typically, company security surveys show that the majority of respondents rank loss or theft as their top mobile-device security concern and second in rank is concern about mobile malware. It is important to remember that while the individual employee may be liable for a device, the company is still liable for the data.

Top Threats Targeting Mobile Devices

According to FCC (2017), the top threats targeting mobile devices are:

- **Data loss**—an employee or hacker accesses sensitive information from device or network. This can be unintentional or malicious, and is considered the biggest threat to mobile devices
- **Social engineering attacks**—a cyber criminal attempts to trick users to disclose sensitive information or install malware. Methods include phishing and targeted attacks
- **Malware**—malicious software that includes traditional computer viruses, computer worms, and Trojan horse programs. Specific examples include the Ikee worm, targeting iOS-based devices, and Pjapps malware that can enroll infected Android devices in a collection of hacker-controlled "zombie" devices known as a "botnet"
- **Data integrity threats**—attempts to corrupt or modify data in order to disrupt operations of a business for financial gain. These can also occur unintentionally
- **Resource abuse**—attempts to misuse network, device, or identity resources. Examples include sending spam from compromised devices or denial-of-service attacks using computing resources of compromised devices
- **Web and network-based attacks**—launched by malicious websites or compromised legitimate sites, these target a device's browser and attempt to install malware or steal confidential data that flows through it

A few simple steps can help to ensure that an organization's information is protected. These include requiring all mobile devices that connect to the business network be equipped with security software and password protection, and providing general security training to make employees aware of the importance of security practices for mobile devices. More specific practices are detailed below (FCC, 2017).

Use security Software on All Smartphones
Security software specifically designed for smartphones can stop hackers and prevent cyber criminals from stealing company information or spying on your employees when they use public networks. It can detect and remove viruses and other mobile threats before they cause problems. It can also eliminate annoying text and multimedia spam messages.

Make Sure All Software Is Up to Date
Mobile devices must be treated like personal computers in that all software on the devices should be kept current, especially the security software. This will protect devices from new variants of malware and viruses that threaten the company's critical information.

Encrypt the Data on Mobile Devices
Business and personal information stored on mobile devices is often sensitive. Encrypting this data is another must. If a device is lost and the Subscriber Identification Module (SIM) card stolen, the thief will not be able to access the data if the proper encryption technology is loaded on the device.

Have Users Password Protect Access to Mobile Devices
In addition to encryption and security updates, it is important to use strong passwords to protect data stored on mobile devices. This will go a long way toward keeping a thief from accessing sensitive data if the device is lost or hacked.

Urge Users to Be Aware of Their Surroundings
Whether entering passwords or viewing sensitive or confidential data, users should exercise caution and be aware of anyone who might be looking over their shoulder.

Employ These Strategies for Email, Texting, and Social Networking
Avoid opening unexpected text messages from unknown senders—as with email, attackers can use text messages to spread malware, phishing scams, and other threats among mobile-device users. The same caution that users have become accustomed to with email should be applied to opening unsolicited text messages.

Don't be lured in by spammers and phishers—to shield business networks from cyber criminals, small businesses should deploy appropriate email security solution, including spam prevention, which protects a company's reputation and manages risks.

Click with caution—just like on stationary PCs, social networking on mobile devices and laptops should be conducted with care and caution. Users should not open unidentified links, chat with unknown people, or visit unfamiliar sites. It doesn't take much for a user to be tricked into compromising a device and the information on it.

Set Reporting Procedures for Lost or Stolen Equipment
In the case of a loss or theft, employees and management should all know what to do next. Processes to deactivate the device and protect its information from intrusion should be in place. Products are also available for the automation of such processes, allowing small businesses to breathe easier after such incidents.

Ensure All Devices Are Wiped Clean Prior to Disposal
Mobile devices have a reset function that allows all data to be wiped. SIM cards should also be removed and destroyed.

EMPLOYEES

Organizations must establish formal recruitment and employment processes to control and preserve the quality of their employees. Many employers have learned the hard way that hiring someone with a criminal record, falsified credentials, or undesirable background can create a legal and financial nightmare.

Experience has shown that without exercising due diligence in hiring, employers run the risk of making unwise hiring choices that can lead to workplace violence, theft, embezzlement, lawsuits for negligent hiring, and numerous other workplace problems.

Develop a Hiring Process that Properly Vets Candidates
The hiring process should be a collaborative effort among different groups of your organization, including recruitment, human resources, security, legal, and management teams. It is important to have a solid application, resume, interview, and reference-checking process to identify potential gaps and issues that may appear in a background check.

Perform Background Checks and Credentialing
Background checks are essential and must be consistent. Using a background screening company is highly recommended. The standard background screening should include the following checks:

- Employment verification
- Education verification
- Criminal records
- Drug testing
- The U.S. Treasury Office of Foreign Affairs and Control
- Sex offender registries
- Social security traces and validation

Depending on the type of your business, other screening criteria may consist of credit checks, civil checks, and federal criminal checks. Conducting post-hire checks for all employees every two to three years, depending on your industry, is also recommended.

Take Care in Dealing with Third Parties

Employers should properly vet partner companies through which your organization hires third-party consultants. To ensure consistent screening criteria are enforced for third-part consultants, you need to explicitly set the credentialing requirements in your service agreement. State in the agreement that the company's credentialing requirements must be followed.

Set Appropriate Access Controls for Employees

Both client data and internal company data are considered confidential and need particular care when viewed, stored, used, transmitted, or disposed. It is important to analyze the role of each employee and set data access control based upon role. If a role does not require the employee to ever use sensitive data, the employee's access to the data should be strictly prohibited. However, if the role requires the employee to work with sensitive data, the level of access must be analyzed thoroughly and be assigned in a controlled and tiered manner following "least-privilege" principles, which allow the employee to only access data that is necessary to perform his or her job.

If the organization does not have a system in place to control data access, the following precautions are strongly recommended. Every employee should (FCC, 2017):

- Never access or view client data without a valid business reason. Access should be on a need-to-know basis.
- Never provide confidential data to anyone—client representatives, business partners, or even other employees—unless you are sure of the identity and authority of that person.
- Never use client data for development, testing, training presentations, or any purpose other than providing production service, client-specific testing, or production diagnostics. Only properly sanitized data that can't be traced to a client, client employee, customer, or the organization's employee should be used for such purposes.
- Always use secure transmission methods such as secure email, secure file transfer (from application to application), and encrypted electronic media (e.g., CDs, USB drives, or tapes).
- Always keep confidential data (hard copy and electronic) only as long as it is needed.
- Follow a "clean desk" policy, keeping workspaces uncluttered and securing sensitive documents so that confidential information does not get into the wrong hands.
- Always use only approved document disposal services or shred all hard-copy documents containing confidential information when finished using them. Similarly, use only approved methods that fully remove all data when disposing of, sending out for repair, or preparing to reuse electronic media.

Provide Security Training for Employees

Security awareness training teaches employees to understand system vulnerabilities and threats to business operations that are present when using a computer on a business network.

A strong IT security program must include training IT users on security policies, procedures, and techniques, as well as the various management, operational, and technical controls necessary and available to keep IT resources secure. In addition, IT infrastructure managers must have the skills necessary to carry out their assigned duties effectively. Failure to give attention to the area of security training puts an enterprise at great risk because security of business resources is as much a human issue as it is a technology issue.

Technology users are the largest audience in any organization and are the single most important group of people who can help to reduce unintentional errors and IT vulnerabilities. Users may include employees, contractors, foreign or domestic guest researchers, other personnel, visitors, guests, and other collaborators or associates requiring access. Users must:

- Understand and comply with security policies and procedures
- Be appropriately trained in the rules of behavior for the systems and applications to which they have access
- Work with management to meet training needs
- Keep software and applications updated with security patches
- Be aware of actions they can take to better protect company information. These actions include: proper password usage, data backup, proper antivirus protection, reporting any suspect incidents or violations of security policy, and following rules established to avoid social engineering attacks and deter the spread of spam or viruses and worms

A clear cataloging of what is considered sensitive data versus non-sensitive data is also needed. More often than not, the following data are considered sensitive information that should be handled with precaution:

- Government-issued identification numbers (e.g., social security numbers and driver's license numbers)
- Financial account information (bank account numbers, credit card numbers, etc.)
- Medical records
- Health insurance information
- Salary information
- Passwords

The training should cover security policies for all means of access and transmission, including secure databases, email, file transfer, encrypted electronic media, and hard copies.

Employers should constantly emphasize the critical nature of data security. Regulatory scheduled refresher training course should be established in order to install the data security culture of the organization. Additionally, distribute news articles related to data privacy and security in employee training, and send organization-wide communication on notable platforms for data-privacy-related news as reminders to employees.

Operational Security

Although operational security, or OPSEC, has its origins in security information important to military operations, it has applications across the business community today.

In a commercial perspective, OPSEC is the process of denying hackers access to any information about the capabilities or intentions of a business by identifying, controlling, and protecting evidence of the planning and execution of activities that are essential for the success of operations.

OPSEC is a continuous process that consists of five distinct actions:

- Identify information that is critical to the business
- Analyze the threat to critical information
- Analyze the vulnerabilities to the business that would allow a cyber criminal to access critical information
- Assess the risks to your business if the vulnerabilities are exploited
- Apply countermeasures to mitigate the risk factors

In addition to being a five-step process, OPSEC is also a mindset that all business employees should embrace. By educating oneself on OPSEC risks and methodologies, protecting sensitive information that is critical to the success of your business becomes second nature.

This section explains the OPSEC process and provides some general guidelines that are applicable to most businesses. An understanding of the following terms is required before the process can be explained:

- *Critical information*—specific data about business strategies and operations that are needed by cyber criminals to hamper or harm businesses from successfully operating
- *OPSEC indicators*—business operations and publicly available information that can be interpreted or pieced together by a cyber criminal to derive critical information
- *OPSEC vulnerability*—a condition in which business operations provide OPSEC indicators that may be obtained and accurately evaluated by a cyber criminal to provide a basis for hampering or harming successful business operations.

Identity of Critical Information

The identification of critical information is important in that it focuses the remainder of the OPSEC process on protecting vital information rather than attempting

IT Security Action Plan

to protect all information relevant to business operations. Given that any business has limited time, personnel, and money for developing secure business practices, it is essential to focus those limited resources on protecting information that is most critical to successful business operations. Examples of critical information include, but should not be limited to, the following (FCC, 2017):

- Customer lists and contact information
- Contracts
- Patents and intellectual property
- Leases and deeds
- Policy manuals
- Articles of incorporation
- Corporate papers
- Laboratory notebooks
- Audio tapes
- Video tapes
- Photographs and slides
- Strategic plans and board meeting minutes

Keep in mind that what is critical information for one business may not be critical for another business. Use the company's mission as a guide for determining what data are truly vital.

Analyze Threats

Analyzing threats involves research and analysis to identify likely cyber criminals who may attempt to obtain crucial information regarding a company's operations. OPSEC planners in businesses should answer the following critical information questions:

- Who might be a cyber criminal (e.g., competitors, politically motivated hackers, etc.)?
- What are the cyber criminal's goals?
- What actions might the cyber criminal take?
- What critical information does the cyber criminal already have on the company's operations (i.e., what is already publicly available)?

Analyze Vulnerabilities

Analyzing to identify the vulnerabilities of a business in protecting critical information is crucial. It requires examining each aspect of security that seeks to protect your critical information and then comparing those indicators with the threats identified in the previous step. Common vulnerabilities for businesses include the following:

- Poorly secured mobile devices that have access to critical information
- Lack of policy on what information and networked equipment can be taken home from work or taken abroad on travel
- Storage of critical information on personal email accounts or other non-company networks

- Lack of policy on what business information can be posted to or accessed by social network sites

Assess Risk

Assessing risk consists of two components. First, OPSEC managers must analyze the vulnerabilities identified in the previous and identify possible OPSEC measures to mitigate each one. Second, specific OPSEC measures must be selected for execution based upon a risk assessment done by the company's senior leadership. Risk assessment requires comparing the estimated cost associated with implementing each possible OPSEC measure to the potential harmful effects on business operations resulting from the exploitation of a particular vulnerability.

OPSEC measure may entail some cost in time, resources, personnel, or interference with normal operations. If the cost to achieve OPSEC protection exceeds the cost of the harm that an intruder could inflict, the application of the measure is inappropriate. Because the decision not to implement a particular OPSEC measure entails risks, this step requires the company's leadership approval.

Payment Cards

If an organization accepts payment by credit or debit cards, it is important to have security steps in place to ensure customer information is safe. A business also may have security obligations pursuant to agreements with the bank or payment services processor. These entities can help companies prevent fraud. In addition, free resources and general security tips are available to learn how to keep sensitive information—beyond payment information—safe.

Understand and Catalog Customer and Card Data You Keep

- Make a list of the type of customer and card information you collect and keep—names, addresses, identification information, payment card numbers, magnetic stripe date, bank account details, and social security numbers. It's not only card numbers criminals want; they're looking for all types of personal information, especially if it helps them commit identity fraud.
- Understand where you keep such information and how it is protected.
- Determine who has access to this data and if they need to have access.

Evaluate Whether the Company Needs to Keep All the Data It Has Stored

- Once an organization knows what information it collects and stores, the company needs to evaluate whether it really needs to keep it. Often businesses may not realize they're logging or otherwise keeping unnecessary data until they conduct an audit. Not keeping sensitive data in storage makes it harder for criminals to steal it.
- If the business has been using card numbers for purposes other than payment transactions, such as a customer loyalty program, the merchant processes should be asked to use alternative data instead. Tokenization, for

IT Security Action Plan

example, is a technology that masks card numbers and replaces it with an alternate number that can't be used for fraud.

Use Secure Tools and Services
- The payments industry maintains lists of hardware, software, and service providers who have been validated against industry security requirements.
- Small businesses that use integrated payment systems, in which the card terminal is connected to a larger computer system, can check the list of validated payment applications to make sure any software they employ has been tested.
- Have a conversation about security with the business provider if the products or services currently being used are not on the lists.

Control Access to Payment Systems
- Whether an organization uses a more complicated payment system or a simple stand-alone terminal, make sure access is controlled.
- Isolate payment systems from other less secure programs, especially those connected to the Internet. For example, don't use the same computer to process payments and surf the Internet.
- Control or limit access to payment systems to only employees who need access.
- Make sure the company uses a secure system for remote access or eliminate remote access if it is not needed so that criminals can't infiltrate the company system from the Internet.

Use Security Tools and Resources
Companies should work with their bank or processor and ask about the anti-fraud measures, tools, and services the company can use to ensure criminals can't use stolen card information at the company.

- For e-commerce retailers:
  - The CVV2 code is the three-digit number on the signature panel that can help verify that the customer has physical possession of the card and not just the account number.
  - Retailers can also use Address Verification Service to ensure that cardholder has provided the correct billing address associated with the account.
  - Services such as Verified by Visa prompt the cardholder to enter a personal password confirming their identity and provide an extra layer of protection.
- For brick and mortar retailers:
  - Swipe the card and get an electronic authorization for the transaction.
  - Check that the signature matches the card.
  - Ensure your payment terminal is secure and safe from tampering.

> **THE SECURITY BASICS**
>
> - Use strong, unique passwords and change them frequently.
> - Use up-to-date firewall and antivirus technologies.
> - Do not click on suspicious links you may receive by email or encounter online.

DIGITAL NETWORK SECURITY

We depend on critical infrastructure every day. Our ability to travel, to communicate with friends and family, to conduct business, to handle our finances, and even our ability to access clean, safe food and water are all reliant upon our Nation's critical infrastructure networks and systems. These essential services that underlie daily life in American society are increasingly being run on digital networks. Every day, people connect to the national grid without even realizing it from their smart phones, computers, and tablets. As a result, these critical systems are prime targets for cyber attacks from those seeking to cause our country harm. DHS (2016)

On April 23, 2000, police in Queensland, Australia, stopped a car on the road and found a stolen computer and radio inside. Using commercially available technology, a disgruntled former employee had turned his vehicle into a pirate command center of sewage treatment along Australia's Sunshine Coast. The former employee's arrest solved a mystery that had troubled the Maroochy Shire wastewater system for two months. Somehow the system was leaking hundreds of thousands of gallons of putrid sewage into parks, rivers and the manicured grounds of a Hyatt Regency hotel—marine life died, the creek water turned black and the stench was unbearable for residents. Until the former employee's capture—during his 46th successful intrusion—the utility's managers did not know why.

Specialists study this case of cyber-terrorism because, at the time, it was the only one known in which someone used a digital control system deliberately to cause harm. The former employee's intrusion shows how easy it is to break in—and how restrained he was with his power.

To sabotage the system, the former employee set the software on his laptop to identify itself as a pumping station, and then suppressed all alarms. The former employee was the "central control station" during his intrusions, with unlimited command of 300 SCADA nodes governing sewage and drinking water alike.

The bottom line: as serious as the former employee's intrusions were they pale in comparison with what he could have done to the fresh water system—he could have done anything he liked (Barton Gellman, 2002).

Other reports of digital network exploits illustrate the debilitating effects such attacks can have on the nation's security, economy, and on public health and safety.

- In May 2015, media sources reported that data belonging to 1.1 million health insurance customers in the Washington, D.C., area were stolen in a

IT Security Action Plan

cyber attack on a private insurance company. Attackers accessed a database containing names, birth dates, email addresses, and subscriber ID numbers of customers.
- In December 2014, the industrial Control Systems Cyber Emergency Response Team (ICS-CERT, which works to reduce risks within and across all critical infrastructure sectors by partnering with law enforce agencies) issued an updated alert on a sophisticated malware campaign compromising numerous industrial control system environments. Their analysis indicated that this campaign had been ongoing since at least 2011.
- In the January–April 2014 issue of its Monitor report, ICS-CERT reported that a public utility had been compromised when a sophisticated threat actor gained unauthorized access to its control system network through a vulnerable remote access capability configured on the system. The incident highlighted the need to evaluate security controls employed at the perimeter and ensure that potential intrusion vectors are configured with appropriate security controls, monitoring, and detection capabilities.
- In December 2016, a Wisconsin couple was charged after the duo allegedly defrauded Enterprise Credit Union in Brookfield of more than $300,000 after one of the defendants, who managed the bank's accounts, had her co-conspirator encash bank checks worth $980 several times each week beginning in May 2015. The charges allege that the couple used the money to buy drugs (http://www.wauwatrosanow.com/story/news/cirme/2016/12/09/two-charged-allegedly-scamming-credit-unions-over-300k/95207718/).

In 2000, the FBI identified and listed threats to critical infrastructure. These threats are listed and described in Table 10.1. In 2015, the Government Accountability Office (GAO) described the sources of digital network-based threats. These threats are listed and described in detail in Table 10.2.

DID YOU KNOW?

Presidential Policy Directive 21 defined "all hazards" as a threat to an incident, natural or man-made, that warrants action to protect life, property, the environment, and public health or safety, and to minimize disruptions of government, social, or economic activities.

Threats to systems supporting critical infrastructure are evolving and growing. As shown in Table 10.2, cyber threats can be unintentional or intentional. Unintentional or non-adversarial threats include equipment failures, software coding errors, and the actions of poorly trained employees. They also include natural disasters and failures of critical infrastructure on which the organization depends but are outside of its control. Intentional threats include both targeted and untargeted attacks form a variety of sources, including criminal groups, hackers, disgruntled employees, foreign

TABLE 10.1
Threats to Digital Networks Observed by the FBI

Threat	Description
Criminal groups	There is an increased use of cyber intrusions by criminal group who attack systems for purposes of monetary gain.
Foreign intelligence services	Foreign intelligence services use cyber tools as part of their information-gathering and espionage activities.
Hackers	Hackers sometimes crack into networks for the thrill of the challenge or for bragging rights in the hacker community. While remote cracking once required a fair amount of skill or computer knowledge, hackers can now download attack scripts and protocols from the Internet and launch them against victim sites. Thus, while attack tools have become more sophisticated, they have also become easier to use.
Hacktivists	Hacktivism refers to politically motivated attacks on publicly accessible Web pages or email servers. These groups and individuals overload email servers and hack into websites to send a political message.
Information warfare	Several nations are aggressively working to develop information warfare doctrine, programs, and capabilities. Such capabilities enable a single entity to have a significant and serious impact by disrupting the supply, communications, and economic infrastructures that support military power—impacts that, according to the Director of Central Intelligence, can affect the daily lives of Americans across the country.
Inside threat	The disgruntled organization insider is a principal source of computer crimes. Insiders may not need a great deal of knowledge about computer intrusions because their knowledge of a victim system often allows them to gain unrestricted access to cause damage to the system or to steal system data. The insider threat also includes outsourcing vendors.
Virus writers	Virus writers are posing an increasingly serious threat. Several destructive computer viruses and "worms" have harmed files and hard drives, including the Melissa Macro Virus, the Explore.Zip worm, the CIH (Chernobyl) Virus, Nimda, and Code Red.

Source: FBI, 2000; 2014.

nation engaged in espionage and information warfare, and terrorists. These threat adversaries vary in terms of the capabilities of the actors, their willingness to act, and their motives, which can include seeking monetary gain or seeking an economic, political, or military advantage (GAO, 2015).

THE BOTTOM LINE

Today's developing "information age" technology has intensified the importance of critical infrastructure protection, in which digital network or cyber security has become as critical as physical security to protecting virtually all critical infrastructure sectors. The Department of Defense (DoD) determined that cyber threats to

TABLE 10.2
Common Digital Network Threat Sources

Source	Description
Non-adversarial, malicious	
Failure in information technology equipment	Failures in displays, sensors, controllers, and information technology hardware responsible for data storage, processing, and communications
Failure in environmental controls	Failures in temperature/humidity controllers or power supplies
Software coding errors	Failures in operating systems, networking, and general-purpose and mission-specific applications
Natural or man-made disaster	Events beyond an entity's control such as fires, floods/tsunamis, tornadoes, hurricanes, and earthquakes
Unusual or natural event	Natural events beyond the entity's control that are not considered to be disasters (e.g., sunspots)
Infrastructure failure or outage	Failure or outage of telecommunications or electrical power
Unintentional user errors	Failures resulting from erroneous, accidental actions taken by individuals (both system users and administrators) in the course of executing their everyday responsibilities
Adversarial	
Hackers or hacktivists	Hackers break networks for the challenge or for revenge, stalking, or monetary gain, among other reasons. Hactivists are ideologically motivated actors who use cyber exploits to further political goals.
Malicious insiders	Insiders (e.g., disgruntled organization employees, including contractors) may not need a great deal of knowledge about computer intrusions because their position with the organization often allows them to gain unrestricted access and cause damage to the target system or to steal system data. These individuals engage in purely malicious activities which should not be confused with non-malicious insider accidents.
Nations	Nations, as well as nation-states, and state-sponsored and state-sanctioned programs use cyber tools as part of their information-gathering and espionage activities. In addition, several nations are aggressively working to develop information warfare doctrine, programs, and capabilities.
Criminal groups and organize crime	Criminal groups seek to attack systems for monetary gain. Specifically, organized criminal groups use cyber exploits to commit identity theft, online fraud, and computer extortion.
Terrorists	Terrorists seek to destroy, incapacitate, or exploit critical infrastructures in order to threaten national security, cause mass casualties, weaken the economy, and damage public morale and confidence.
Unknown malicious outsiders	Unknown malicious outsiders are threat sources or agents that, due to a lack of information, agencies are unable to classify as being one of the five types of threat sources or agents listed above.

Source: GAO analysis of unclassified government and nongovernmental data. GAO 16–79.

contractors' unclassified information systems represented an unacceptable risk of compromise to DoD information and posed a significant risk to U.S. national security and economic security interests.

In the past few years, especially since 9/11, it has been somewhat routine for us to pick up a newspaper, magazine, or view a television news program where a major topic of discussion is cyber security or its lack thereof. For example, recently there has been discussion about Russian hackers trying to influence the U.S. 2016 elections. Many of the cyber intrusion incidents we read or hear about have added new terms or new uses for old terms to our vocabulary. For example, old terms such as botnets (short for robot networks, also balled bots, zombies, botnet fleets, and many others) which are groups of computers that have been compromised with malware such as Trojan horses, worms, backdoors, remote control software, and viruses have taken on new connotations in regard to cyber security issues. Relatively new terms such as scanners, Windows NT hacking tools, ICQ hacking tools, mail bombs, sniffer, logic bomb, nukers, dots, backdoor Trojan, key loggers, hackers' Swiss knife, password crackers, blended threats, Warhol worms, flash threats, targeted attacks, and BIOS crackers are now commonly read or heard. New terms have evolved along with various control mechanisms. For example, because many control systems are vulnerable to attacks of varying degrees, these attack attempts range from telephone line sweeps (wardialing), to wireless network sniffing (wardriving), to physical network port scanning, to physical monitoring and intrusion. When wireless network sniffing is performed at (or near) the target point by a pedestrian (warwalking), meaning that instead of a person being in an automotive vehicle, the potential intruder sniffing the network for weaknesses or vulnerabilities may be on foot, posing as a person walking, but they may have a handheld personal digital assistant (PDA) device or laptop computer (Warwalking 2003). Further, adversaries can leverage common computer software programs, such as Adobe Acrobat and Microsoft Office, to deliver a threat by embedding exploits within software files that can be activated when a user opens a file within its corresponding program. Finally, the communications infrastructure and the utilities are extremely dependent on the information technology sector. This dependency is due to the reliance of the communications systems on the software that runs the control mechanism of the operating systems, the management software, the billing software, and any number of other software packages used by industry. Table 10.3 provides descriptions of common exploits or techniques, tactics, and practices used by digital/cyber adversaries.

Not all relatively new and universally recognizable information technology and digital or cyber terms have sinister connotations or meaning, of course. Consider, for example, the following digital terms: backup, binary, bit byte, CD-ROM, CPU, database, email, HTML, icon, memory, cyberspace, modem, monitor, network, RAM, Wi-Fi (wireless fidelity), record, software, and World Wide Web—none of these terms normally generate thoughts of terrorism in most of us.

TABLE 10.3
Common Methods of Digital/Cyber Exploits

Exploit	Description
Watering hole	A method by which threat actors exploit the vulnerabilities of carefully selected websites frequented by users of the targeted system. Malware is then injected to the targeted system via the compromised websites
Phishing and spear phishing	A digital form of social engineering that uses authentic-looking emails, websites, or instant messages to get users to download malware, open malicious attachments, or open links that direct then to a website that requires information or executes malicious code
Insufficient authentication requirements	An exploit that takes advantage of a system's insufficient user authentication and/or any elements of cyber security supporting it to include not limiting the number of failed login attempts, the use of hard-coded credentials, and the use of a broken or risky cryptographic algorithm
Trusted third parties	An exploit that takes advantage of the security vulnerabilities of trusted third parties to gain access to an otherwise secure system
Classic buffer overflow	An exploit that involves the intentional transmission of more data than a program's input buffer can hold, leading to the deletion of critical data and subsequent execution of malicious code
Cryptographic weakness	An exploit that takes advantage of a network employing insufficient encryption when either storing or transmitting data, enabling adversaries to read and/or modify the data stream
Structured Query Language (SQL) injection	An exploit that involves the alteration of a database search in a web-based application, which can be used to obtain unauthorized across to sensitive information in a database, resulting in data loss due to corruption, denial of service, or complete host takeover
Operating system command injection	An exploit that takes advantage of a system's inability to properly neutralize special elements used in operating system commands, allowing the adversaries to execute unexpected commands on the system by either modifying already evoked commands or evoking their own
Cross-site scripting	An exploit that uses third-party web resources to run lines of programming code (referred to as scripts) within the victim's web browser or scriptable application. This occurs when a user, using a browser, visits a malicious website or clicks a malicious link. The most dangerous consequences can occur when this method is used to exploit additional vulnerabilities that may permit an adversary to steal cookies (data exchanged between a web server and a browser), log key stokes, capture screen shots, discover and collect network information, or remotely access and control the victim's machine
Cross-site request forgery	An exploit that takes advantage of an application that cannot, or does not, sufficiently verify whether a well-formed, valid, consistent request was intentionally provided by the user who submitted the request, tricking the victim into executing a falsified request that results in the system or data being compromised.

(Continued)

TABLE 10.3 (CONTINUED)
Common Methods of Digital/Cyber Exploits

Exploit	Description
Path traversal	An exploit that seeks to gain access to files outside of a restricted directory by modifying the directory pathname in an application that does not properly neutralize special elements (e.g., "…," "/," "…/," etc.)
Integer overflow	An exploit where malicious code is inserted that leads to unexpected integer overflow, or wraparound, which can be used by adversaries to control looping or make security decisions in order to cause program crashes, memory corruption, or the execution of arbitrary code via buffer overflow
Uncontrolled format string	Adversaries manipulate externally controlled format strings in print-style functions to gain access to information and/or execute unauthorized code or commands
Open redirect	An exploit where the victim is tricked into selecting a URL (website location) that has been modified to direct them to an external, malicious site which may contain malware that can compromises the victim's machine
Heap-based buffer overflow	Similar to classic buffer overflow, but the buffer that is overwritten is allocated in the heap portion of memory, generally meaning that the buffer was allocated using a memory allocation routine, such as "malloc ( )"
Unrestricted upload of files	An exploit that takes advantage of insufficient upload restrictions, enabling adversaries to upload malware (e.g., .php) in place of the intended file type (e.g., .jpg)
Inclusion of functionality from an un-trusted sphere	An exploit that uses trusted, third-party executable functionality (e.g., web widget or library) as a means of executing malicious code in software whose protection mechanism is unable to determine whether the functionality is from a trusted source, is modified in transit, or whether it is being spoofed
Certificate and certificate authority compromise	Exploits facilitated via the issuance of fraudulent digital certificates (e.g., transport layer security and Secure Socket Layer). Adversaries use these certificates to establish secure connections with the target organization or individual by mimicking a trusted third party
Hybrid of others	An exploit which combines elements of two or more of the aforementioned techniques

Source: GAO (2015).

REFERENCES AND RECOMMENDED READING

Associated Press (AP). 2009. Goal: Disrupt. *The Virginian-Pilot*, Norfolk, VA, 04 April 2009.

DHS. 2003. *The National Strategy for the Physical Protection of Critical Infrastructures and Key Assets.* Accessed 21 December 2019 @ https://dhs.gov/xlibrary/assets/physical_Strat.

DHS. 2009. *National Infrastructure Protection Plan.* Retrieved 05 November 2019 @ http://dhs.gov/xlibrary/assets/NIPP.Plan.pdf.

DHS. 2013. *Homeland Security Directive 7: Critical Infrastructure Identification, Prioritization, and Protection.* Accessed 21 December 2019 @ https://dihs.gov/xabout/laws/gc_1214597989952.shtm.

DHS. 2016. *Information Technology Sector—an Annex to the National Infrastructure Protection Plan.* Washington, DC: Department of Homeland Security.
DOE. 2001. *21 Steps to Improve Cyber Security of SCADA Networks.* Washington, DC: Department of Energy.
FBI. 2000. *Threat to Critical Infrastructure.* Washington, DC: Federal Bureau of Investigation.
FBI. 2007. *Ninth Annual Computer Crime and Security Survey.* FBI: Computer Crime Institute and Federal Bureau of Investigations, Washington, DC.
FBI. 2014. *Protecting Critical Infrastructure and the Importance of Partnerships.* Accessed 21 December 2019 @ https://fib.gov/news/speeches/protecitng-critical-infrastruce-and t-the-importantce-o.
FCC. 2017. *Cyber Security Planning Guide.* Washington, DC: Federal Communication Commission.
FEMA. 2008. *FEMA452: Risk Assessment: A How to Guide.* Accessed 05 January 2008 @ fema.gov/library/file?type=published/filetofile.
FEMA. 2015. *Protecting Critical Infrastructure Against Insider Threats.* Accessed 17 April 2015 @ http://emilms.fema.gov/IS0915/IABsummary.htm.
GAO. 2003. *Critical Infrastructure Protection: Challenges in Securing Control System.* Washington, DC: United States General Accounting Office.
GAO. 2015. *Critical Infrastructure Protection: Sector-Specific Agencies Need to Better Measure Cybersecurity Progress.* Washington, DC: United States Government Accountability Office.
Gellman, B. 2002. Cyber-Attacks by Al Qaeda Feared: Terrorists at Threshold of Using Internet as Tool of Bloodshed, Experts Say. *Washington Post,* June 27, p. A01.
Minter, J.G. 1996. Prevention Chemical Accidents Still a Challenge. *Occupational Hazards,* September, Vol xi 22–24.
National Infrastructure Advisory Council. 2008. *First Report and Recommendations on the Insider Threat to Critical Infrastructure.* Department of Homeland Security, Washington, DC.
NIPC. 2002. *National Infrastructure Protection Center Report.* Washington, DC: National Infrastructure Protection Center.
Spellman, F.R. 1997. *A Guide to Compliance for PSM/RMP.* Lancaster, PA: Technomic Publishing Company.
Stamp, J., et al. 2003. *Common Vulnerabilities in Critical Infrastructure Control Systems.* 2nd ed. Sandia National Laboratories.
US DOE. 2002. *Vulnerability Assessment Methodology: Electric Power Infrastructure.* Department of Energy, Washington, DC.
USEPA. 2005. EPA Needs to Determine What Barriers Prevent Water Systems from Securing Known SCADA Vulnerabilities. In: Harris, J. (ed.), *Final Briefing Report.* Washington, DC: USEPA, p.17–21.
Warwalking. 2003. Warwalking Accessed 05 September 2019 @ http://warwalking.tribe.net.

11 Plant Security

You may say Homeland Security is a Y2K problem that doesn't end Jan. 1 of any given year.

Governor Tom Ridge

Worldwide conflicts are ongoing and seem never ending. One of the most important conflicts of our time, such as the on-going Israeli–Palestinian conflict, is in fact conflict over scarce but vital water resources. This conflict over water, unfortunately, may be a harbinger of things to come.

U.S. Sees Increase in Cyber Attacks on Infrastructure

The top U.S. military official responsible for defending the United States against cyber attacks said Thursday that there had been a 17-fold increase in computer attacks on U.S. infrastructure between 2009 and 2011, initiated by criminal gangs, hackers and other nations.

New York Times, **July 27, 2012**

INTRODUCTION

According to the USEPA (2004), there are approximately 160,000 public water systems (PWSs) in the United States, each of which regularly supplies drinking water to at least 25 persons or 15 service connections. Eighty-four percent of the total U.S. population is served by PWSs, while the remainder is served primarily by private wells. PWSs are divided into community water systems (CWSs) and noncommunity water systems (NCWSs). Examples of CWSs include municipal water systems that serve mobile home parks of residential developments. Examples of NCWSs include schools, factories, churches, commercial campgrounds, hotels, and restaurants.

> **DID YOU KNOW?**
>
> As of 2003, community water systems serve by far the largest proportion of the U.S. population—273 million out of a total population of 290 million (USEPA 2004).

Because drinking water is consumed directly, health effects associated with contamination have long been major concerns. In addition, interruption or cessation of the drinking water supply can disrupt society, impacting human health and critical activities such as fire protection. Although they have no clue as to its true economic value and to its future worth, the general public correctly perceives drinking water as central to the life of an individual and of society. However, the general public knows even less about the importance of wastewater treatment and the fate of its end product.

Wastewater treatment is important for preventing disease and protecting the environment. Wastewater is treated by publicly owned treatment works (POTWs) and by private facilities such as industrial plants. There are approximately 2.3 million miles of distribution system pipes and approximately 16,255 POTWs in the United States. Seventy-five percent of the total U.S. population is served by POTWs, with existing flow rates of less than 1 MGD being considered insufficient; they number approximately 13,057 systems. In terms of population served, 1 MGD equals approximately 10,000 persons served.

Disruption of a wastewater treatment system or service can cause loss of life, economic impacts, and severe public health incidents. If structural damage occurs, wastewater systems can become vulnerable to inadequate treatment. The public is much less sensitive to wastewater as an area of vulnerability than it is to drinking water; however, wastewater systems do provide opportunities for terrorist threats.

Federal and state agencies have long been active in addressing these risks and threats to water and wastewater utilities through regulations, technical assistance, research, and outreach programs. As a result, an extensive system of regulations governing maximum contaminant levels (MCLs) of 90 conventional contaminants (most established by EPA), construction and operating standards (implemented mostly by the states), monitoring, emergency response planning, training, research, and education have been developed to better protect the nation's drinking water supply and receiving waters. Since the events of 9/11, EPA has been designated as the sector-specific agency responsible for infrastructure protection activities for the nation's drinking water and wastewater system. EPA is utilizing its position within the water sector and working with its stakeholders to provide information to help protect the nation's drinking water supply from terrorism or other intentional acts.

CONSEQUENCES OF 9/11

One consequence of the events of September 11, 2001, was the USEPA's directive to establish a Water Protection Task Force to ensure that activities to protect and secure water supply and wastewater treatment infrastructure are comprehensive and carried out expeditiously. Another consequence is a heightened concern among citizens in the United States over the security of their critical water and wastewater infrastructure. The nation's water and wastewater infrastructure, consisting of several thousand publicly owned water/wastewater treatment plants, more than 100,000 pumping stations, hundreds of thousands of miles of water distribution and sanitary sewers, and another 200,000 miles of storm sewers, is one of America's most valuable resources, with treatment and distribution/collection systems valued at more than $2.5 trillion. Wastewater treatment operations taken alone include the sanitary and storm sewers forming an extensive network that runs near or beneath key buildings and roads, and is contiguous to many communication and transportation networks. Significant damage to the nation's wastewater facilities or collection systems would result in: loss of life; catastrophic environmental damage to rivers, lakes,

Plant Security 187

and wetlands; contamination of drinking water supplies; long-term public health impacts; destruction of fish and shellfish production; and disruption to commerce, the economy, and our normal way of life.

Governor Tom Ridge (Henry, 2002) pointed out the security role played by the public professional (I interpret this to include water and wastewater professionals):

> Americans should find comfort in knowing that millions of their fellow citizens are working every day to ensure our security at every level—federal, state, county, municipal. These are dedicated professionals who are good at what they do. I've seen it up close, as Governor of Pennsylvania. ... But there may be gaps in the system. The job of the Office of Homeland Security will be to identify those gaps and work to close them.

It is to shore up the "gaps in the system" that many water and wastewater facilities have been driven to increase security. Moreover, the USEPA in its *Water Protection Task Force Alert #IV: What Wastewater Utilities Can Do Now to Guard Against Terrorist and Security Threats* (October 24, 2001) made several recommendations to increase security and reduce threats from terrorism. The recommendations include:

1. Guarding Against Unplanned Physical Intrusion (Water/Wastewater)
   a. Lock all doors and set alarms at your office, pumping stations, treatment plants, and vaults, and make it a rule that doors are locked and alarms are set.
   b. Limit access to facilities and control access to pumping stations, chemical and fuel storage areas, giving close scrutiny to visitors and contractors.
   c. Post guards at treatment plants, and post "Employee Only" signs in restricted areas.
   d. Control access to storm sewers.
   e. Secure hatches, metering vaults, manholes, and other access points to the sanitary collection system.
   f. Increase lighting in parking lots, treatment bays, and other areas with limited staffing.
   g. Control access to computer networks and control systems, and change the passwords frequently.
   h. Do not leave keys in equipment or vehicles at any time.
2. Making Security a Priority for Employees
   a. Conduct background security checks on employees during hiring and periodically thereafter.
   b. Develop a security program with written plans and train employees frequently.
   c. Ensure all employees are aware of communications protocols with relevant law enforcement, public health, environmental protection, and emergency response organizations.
   d. Ensure that employees are fully aware of the importance of vigilance and the seriousness of breaches in security, and make note of

unaccompanied strangers on the site and immediately notify designated security officers or local law enforcement agencies.
   e. Consider varying the timing of operational procedures if possible so if someone is watching the pattern changes.
   f. Upon the dismissal of an employee, change passcodes and make sure keys and access cards are returned.
   g. Provide customer service staff with training and checklists to handle a threat if it is called in.
3. Coordinating Actions for Effective Emergency Response
   a. Review existing emergency response plans, and ensure they are current and relevant.
   b. Make sure employees have necessary training in emergency operating procedures.
   c. Develop clear protocols and chains of command for reporting and responding to threats along with relevant emergency, law enforcement, environmental, and public health officials, consumers, and the media. Practice the emergency protocols regularly.
   d. Ensure key utility personnel (both on and off duty) have access to crucial telephone numbers and contact information at all times. Keep the call list up to date.
   e. Develop close relationships with local law enforcement agencies and make sure they know where critical assets are located. Request they add your facilities to their routine rounds.
   f. Work with local industries to ensure that their pretreatment facilities are secure.
   g. Report to county or State health officials any illness among the employees that might be associated with wastewater contamination.
   h. Report criminal threats, suspicious behavior, or attacks on wastewater utilities immediately to law enforcement officials and the relevant field office of the Federal Bureau of Investigation.
4. Investing in Security and Infrastructure Improvements
   a. Assess the vulnerability of collection/distribution system, major pumping stations, water and wastewater treatment plants, chemical and fuel storage areas, outfall pipes, and other key infrastructure elements.
   b. Assess the vulnerability of the storm water collection system. Determine where large pipes run near or beneath government buildings, banks, commercial districts, industrial facilities, or are contiguous with major communication and transportation networks.
   c. Move as quickly as possible with the most obvious and cost-effective physical improvements, such as perimeter fences, security lighting, tamper-proofing manhole covers and valve boxes, etc.
   d. Improve computer system and remote operational security.
   e. Use local citizen watches.
   f. Seek financing for more expensive and comprehensive system improvements.

Plant Security

Ideally, in a perfect world, water and wastewater infrastructure would be secured in a layered fashion (aka the barrier approach). Layered security systems are vital. Using the protection "in depth" principle, requiring that an adversary defeat several protective barriers or security layers to accomplish its goal, water and wastewater infrastructure can be made more secure. Protection in depth is a term commonly used by the military to describe security measures that reinforce one another, masking the defense mechanisms from view of intruders, and allowing the defender time to respond to intrusion or attack.

A prime example of the use of the multi-barrier approach to ensure security and safety is demonstrated by the practices of the bottled water industry. In the aftermath of 9/11 and the increased emphasis on Homeland Security, a shifted paradigm of national security and vulnerability awareness has emerged. Recall that in the immediate aftermath of the 9/11 tragedies, emergency responders and others responded quickly and worked to exhaustion. In addition to the emergency responders, bottled water companies responded immediately by donating several million bottles of water to the crews at the crash sites in New York, at the Pentagon, and in Pennsylvania. International Bottled Water Association reports that "within hours of the first attack, bottled water was delivered where it mattered most; to emergency personnel on the scene who required ample water to stay hydrated as they worked to rescue victims and clean up debris" (IBWA 2004, p. 2).

Bottled water companies continued to provide bottled water to responders and rescuers at the 9/11 sites throughout the post-event(s) process(es). These patriotic acts by the bottled water companies, however, beg the question: How do we ensure the safety and security of bottled water provided to anyone? IBWA (2004) has the answer: Using a multi-barrier approach, along with other principles, will enhance the safety and security of bottled water. IBWA (2004, p. 3) described its multi-barrier approach as follows:

> A multi-barrier approach—Bottled water products are produced utilizing a multi-barrier approach, from source to finished product, that helps prevent possible harmful contaminants (physical, chemical or microbiological) from adulterating the finished product as well as storage, production, and transportation equipment. Measures in a multi-barrier approach may include source protection, source monitoring, reverse osmosis, distillation, filtration, ozonation or ultraviolet (UV) light. Many of the steps in a multi-barrier system may be effective in safeguarding bottled water from microbiological and other contamination. Piping in and out of plants, as well as storage silos and water tankers are also protected and maintained through sanitation procedures. In addition, bottled water products are bottled in a controlled, sanitary environment to prevent contamination during the filling operation.

In water and wastewater infrastructure security, protection in depth is used to describe a layered security approach. A protection in depth strategy uses several forms of security techniques and/or devices against an intruder and does not rely on one single defense mechanism to protect infrastructure. By implementing multiple layers of security, a hole or flaw in one layer is covered by the other layers. An intruder will have to intrude through each layer without being detected in the process—the layered approach implies that no matter how an intruder attempts to accomplish his goal, he will encounter effective elements of the physical protection system.

In the following sections, various security hardware and/or devices are described. These devices serve the main purpose of providing security against physical and/or digital intrusion. That is, they are designed to delay and deny intrusion and are normally coupled with detection and assessment technology. Keep in mind, however, when it comes to making "anything" absolutely secure from intrusion or attack, there is inherently, or otherwise, no silver bullet.

SECURITY HARDWARE DEVICES

The USEPA (2005) groups the water/wastewater infrastructure security devices or products into four general categories:

- Physical asset monitoring and control devices
- Water monitoring devices
- Communication/integration
- Cyber protection devices

Physical Asset Monitoring and Control Devices

Aboveground, Outdoor Equipment Enclosures

Water and wastewater systems consist of multiple components spread over a wide area, and typically include a centralized treatment plant, as well as distribution or collection system components that are typically distributed at multiple locations throughout the community. However, in recent years, distribution and collection system designers have favored placing critical equipment—especially assets that require regular use and maintenance—aboveground.

One of the primary reasons for doing so is that locating this equipment aboveground eliminates the safety risks associated with confined space entry, which is often required for the maintenance of equipment located belowground. In addition, space restrictions often limit the amount of equipment that can be located inside, and there are concerns that some types of equipment (such as backflow prevention devices) can, under certain circumstances, discharge water that could flood pits, vaults, or equipment rooms. Therefore, many pieces of critical equipment are located outdoors and aboveground. Many different system components can be installed outdoors and aboveground. Examples of these types of components could include:

- Backflow prevention devices
- Air release and control valves
- Pressure vacuum breakers
- Pumps and motors
- Chemical storage and feed equipment
- Meters
- Sampling equipment
- Instrumentation

Much of this equipment is installed in remote locations and/or in areas where the public can access it.

One of the most effective security measures for protecting aboveground equipment is to place them inside a building. When/where this is not possible, enclosing the equipment or parts of the equipment using some sort of commercial or homemade add-on structure may help to prevent tampering with the equipment. Equipment enclosures can generally be categorized into one of four main configurations, which include:

- One piece, drop over enclosures
- Hinged or removable top enclosures
- Sectional enclosures
- Shelters with access locks

Other security features that can be implemented on aboveground, outdoor equipment enclosures include locks, mounting brackets, tamper-resistant doors, and exterior lighting.

Alarms

An *alarm system* is a type of electronic monitoring system that is used to detect and respond to specific types of events—such as unauthorized access to an asset, or a possible fire. In water and wastewater systems, alarms are also used to alert operators when process operating or monitoring conditions go out of preset parameters (i.e., process alarms). These types of alarms are primarily integrated with process monitoring and reporting systems (i.e., SCADA systems). Note that this discussion does not focus on alarm systems that are not related to a utility's processes.

Alarm systems can be integrated with fire detection systems, IDSs, access control systems, or closed circuit television (CCTV) systems, such that these systems automatically respond when the alarm is triggered. For example, a smoke detector alarm can be set up to automatically notify the fire department when smoke is detected, or an intrusion alarm can automatically trigger cameras to turn on in a remote location so that personnel can monitor that location.

An alarm system consists of sensors that detect different types of events; an arming station that is used to turn the system on and off; a control panel that receives information, processes it, and transmits the alarm; and an annunciator that generates a visual and/or audible response to the alarm. When a sensor is tripped it sends a signal to a control panel, which triggers a visual or audible alarm and/or notifies a central monitoring station. A more complete description of each of the components of an alarm system is provided below.

Detection devices (also called *sensors*) are designed to detect a specific type of event (such as smoke, intrusion, etc.). Depending on the type of event they are designed to detect, sensors can be located inside or outside of the facility or other assets. When an event is detected, the sensors use some type of communication method (such as wireless radio transmitters, conductors, or cables) to send signals to the control panel to generate the alarm. For example, a smoke detector sends a signal to a control panel when it detects smoke.

An *arming station*, which is the main user interface with the security system, allows the user to arm (turn on), disarm (turn off), and communicate with the system. How a specific system is armed will depend on how it is used. For example, while IDSs can be armed for continuous operation (24 hours/day), they are usually armed and disarmed according to the work schedule at a specific location so that personnel going about their daily activities do not set off the alarms. In contrast, fire protection systems are typically armed 24 hours/day.

The *control panel* receives information from the sensors and sends it to an appropriate location, such as to a central operations station or to a 24-hour monitoring facility. Once the alarm signal is received at the central monitoring location, personnel monitoring for alarms can respond (such as by sending security teams to investigate or by dispatching the fire department).

The *annunciator* responds to the detection of an event by emitting a signal. This signal may be visual, audible, electronic, or a combination of these three. For example, fire alarm signals will always be connected to audible annunciators, whereas intrusion alarms may not be.

Alarms can be reported locally, remotely, or both. A *local alarm* emits a signal at the location of the event (typically using a bell or siren). A "local only" alarm emits a signal at the location of the event but does not transmit the alarm signal to any other location (i.e., it does not transmit the alarm to a central monitoring location). Typically, the purpose of a "local only" alarm is to frighten away intruders, and possibly to attract the attention of someone who might notify the proper authorities. Because no signal is sent to a central monitoring location, personnel can only respond to a local alarm if they are in the area and can hear and/or see the alarm signal.

Fire alarm systems must have local alarms, including both audible and visual signals. Most fire alarm signal and response requirements are codified in the National Fire Alarm Code, National Fire Protection Association (NFPA) 72. NFPA 72 discusses the application, installation, performance, and maintenance of protective signaling systems and their components. In contrast to fire alarms, which require a local signal when fire is detected, many IDSs do not have a local alert device, because monitoring personnel do not wish to inform potential intruders that they have been detected. Instead, these types of systems silently alert monitoring personnel that an intrusion has been detected, thus allowing monitoring personnel to respond.

In contrast to systems that are set up to transmit "local only" alarms when the sensors are triggered, systems can also be set up to transmit signals to a *central location*, such as to a control room or guard post at the utility, or to a police or fire station. Most fire/smoke alarms are set up to signal both at the location of the event and at a fire station or central monitoring station. Many insurance companies require that facilities install certified systems that include alarm communication to a central station. For example, systems certified by the Underwriters Laboratory (UL) require that the alarm be reported to a central monitoring station.

The main difference between alarm systems lies in the types of event detection devices used in different systems. *Intrusion sensors*, for example, consist of two main categories: perimeter sensors and interior (space) sensors. *Perimeter intrusion sensors* are typically applied on fences, doors, walls, windows, etc., and are designed to detect an intruder before he/she accesses a protected asset (i.e., perimeter intrusion sensors are used to detect intruders attempting to enter through a door, window, etc.). In contrast, *interior intrusion sensors* are designed to detect an intruder who has already accessed the protected asset (i.e., interior intrusion sensors are used to detect intruders once they are already within a protected room or building). These two types of detection devices can be complementary, and they are often used together to enhance security for an asset. For example, a typical intrusion alarm system might employ a perimeter glass-break detector that protects against intruders accessing a room through a window, as well as an ultrasonic interior sensor that detects intruders that have gotten into the room without using the window.

Fire detection/fire alarm systems consist of different types of fire detection devices and fire alarm systems available. These systems may detect fire, heat, smoke, or a combination of any of these. For example, a typical fire alarm system might consist of heat sensors, which are located throughout a facility and which detect high temperatures or a certain change in temperature over a fixed time period. A different system might be outfitted with both smoke and heat detection devices.

When a sensor in an alarm system detects an event, it must communicate an alarm signal. The two basic types of alarm communication systems are hardwired and wireless. Hardwired systems rely on wire that is run from the control panel to each of the detection devices and annunciators. Wireless systems transmit signals from a transmitter to a receiver through the air—primarily using radio or other waves. Hardwired systems are usually lower cost, more reliable (they are not affected by terrain or environmental factors), and significantly easier to troubleshoot than wireless systems. However, a major disadvantage of hardwired systems is that it may not be possible to hardwire all locations (for example, it may be difficult to hardwire remote locations). In addition, running wires to their required locations can be both time consuming and costly. The major advantage to using wireless systems is that they can often be installed in areas where hardwired systems are not feasible. However, wireless components can be much more expensive when compared to hardwired systems. In addition, in the past, it has been difficult to perform self-diagnostics on wireless systems to confirm that they are communicating properly with the controller. Presently, the majority of wireless systems incorporate supervising circuitry, which allows the subscriber to know immediately if there is a problem with the system (such as a broken detection device or a low battery), or if a protected door or window has been left open.

Backflow Prevention Devices

As their name suggests, backflow prevention devices are designed to prevent backflow, which is the reversal of the normal and intended direction of water flow in a water system. Backflow is a potential problem in a water system because it can spread contaminated water back through a distribution system. For example, backflow at uncontrolled cross-connections (cross-connections are any actual or potential connection between the public water supply and a source of contamination) or pollution can allow pollutants or contaminants to enter the potable water system. More specifically, backflow from private plumbing systems, industrial areas, hospitals, and other hazardous contaminant-containing systems, into public water mains and wells poses serious public health risks and security problems. Cross-contamination from private plumbing systems can lead to biological hazards (due to bacteria or viruses) or transfer of toxic substances that can contaminate and sicken an entire population in the event of backflow. The majority of historical incidences of backflow have been accidental, but growing concern that contaminants could be intentionally backfed into a system is prompting increased awareness among private homes, businesses, industries, and areas most vulnerable to intentional strikes. Therefore, backflow prevention is a major tool for the protection of water systems.

Backflow may occur under two types of conditions: backpressure and backsiphonage. *Backpressure* is the reverse from normal flow direction within a piping system that is the result of the downstream pressure being higher than the supply pressure. These reductions in the supply pressure occur whenever the amount of water being used exceeds the amount of water supplied, such as during water line flushing, firefighting, or breaks in water mains. *Backsiphonage* is the reverse from normal flow direction within a piping system that is caused by negative pressure in the supply piping (i.e., the reversal of normal flow in a system caused by a vacuum or partial vacuum within the water supply piping). Backsiphonage can occur where there is a high velocity in a pipeline; when there is a line repair or break that is lower than a service point; or when there is lowered main pressure due to high water withdrawal rate, such as during firefighting or water main flushing.

To prevent backflow, various types of backflow preventers are appropriate for use. The primary types of backflow preventers are:

- Air gap drains
- Double check valves
- Reduced pressure principle assemblies
- Pressure vacuum breakers

Barriers

Active Security Barriers (Crash Barriers)

Active security barriers (also known as crash barriers) are large structures that are placed in roadways at the entry and exit points to protected facilities to control vehicle access to these areas. These barriers are placed perpendicular to traffic to block the roadway, so that the only way that traffic can pass the barrier is for the barrier to be moved out of the roadway. These types of barriers are typically constructed from sturdy materials, such as concrete or steel, such that vehicles cannot penetrate through them. They are also designed at a certain height off the roadway so that vehicles cannot go over them.

The key difference between active security barriers, which include wedges, crash beams, gates, retractable bollards, and portable barricades, and passive security barriers, which include non-movable bollards, jersey barriers, and planters, is that active security barriers are designed so that they can be raised and lowered or moved out of the roadway easily to allow authorized vehicles to pass them. Many of these types of barriers are designed so that they can be opened and closed automatically (e.g., mechanized gates and hydraulic wedge barriers), while others are easy to open and close manually (e.g., swing crash beams and manual gates). In contrast to active

Plant Security

barriers, passive barriers are permanent, non-movable barriers, and thus they are typically used to protect the perimeter of a protected facility, such as sidewalks and other areas that do not require vehicular traffic to pass them. Several of the major types of active security barriers such as wedge barriers, crash beams, gates, bollards, and portable/removable barricades are described below.

Wedge barriers are plated, rectangular steel buttresses approximately 2–3 feet high that can be raised and lowered from the roadway. When they are in the open position, they are flush with the roadway and vehicles can pass over them. However, when they are in the closed (armed) position, they project up from the road at a 45-degree angle, with the upper end pointing toward the oncoming vehicle and the base of the barrier away from the vehicle. Generally, wedge barriers are constructed from heavy gauge steel, or concrete that contains an impact-dampening iron rebar core that is strong and resistant to breaking or cracking, thereby allowing them to withstand the impact from a vehicle attempting to crash through them. In addition, both of these materials help to transfer the energy of the impact over the barrier's entire volume, thus helping to prevent the barrier from being sheared off its base. In addition, because the barrier is angled away from traffic, the force of any vehicle impacting the barrier is distributed over the entire surface of the barrier and is not concentrated at the base, which helps prevent the barrier from breaking off at the base. Finally, the angle of the barrier helps stop any vehicles attempting to drive over it.

Wedge barriers can be fixed or portable. Fixed wedge barriers can be mounted on the surface of the roadway ("surface-mounted wedges") or in a shallow mount in the road's surface, or they can be installed completely below the road surface. Surface-mounted wedge barricades operate by rising from a flat position on the surface of the roadway, while shallow-mount wedge barriers rise from their resting position just below the road surface. In contrast, below-surface wedge barriers operate by rising from beneath the road surface. Both the shallow-mounted and surface-mounted barriers require little or no excavation, and thus do not interfere with buried utilities. All three barrier mounting types project above the road surface and block traffic when they are raised into the armed position. Once they are disarmed and lowered, they are flush with the road, thereby allowing traffic to pass portable wedge barriers that are moved into place on wheels that are removed after the barrier has been set into place.

Installing rising wedge barriers requires preparation of the road surface. Installing surface-mounted wedges does not require that the road be excavated; however, the road surface must be intact and strong enough to allow the bolts anchoring the wedge to the road surface to attach properly. Shallow-mount and below-surface wedge barricades require excavation of a pit that is large enough to accommodate the wedge structure, as well as any arming/disarming mechanisms. Generally, the bottom of the excavation pit is lined with gravel to allow for drainage. Areas not sheltered from rain or surface runoff can install a gravity drain or self-priming pump.

Crash beam barriers consist of aluminum beams that can be opened or closed across the roadway. While there are several different crash beam designs, every crash beam system consists of an aluminum beam that is supported on each side by a solid footing or buttress, which is typically constructed from concrete, steel, or some other strong material. Beams typically contain an interior steel cable (typically at least 1 inch in diameter) to give the beam added strength and rigidity. The beam is connected by a heavy-duty hinge or a similar mechanism to one of the footings so that it can swing or rotate out of the roadway when it is open, and can swing back across the road when it is in the closed (armed) position, blocking the road and inhibiting access by unauthorized vehicles. The non-hinged end of the beam can be locked into its footing, thus providing anchoring for the beam on both sides of the road and increasing the beam's resistance to any vehicles attempting to penetrate through it. In addition, if the crash beam is hit by a vehicle, the aluminum beam transfers the impact energy to the interior cable, which in turn transfers the impact energy through the footings and into their foundation, thereby minimizing the chance that the impact will snap the beam and allow the intruding vehicle to pass through.

Crash beam barriers can employ drop-arm, cantilever, or swing beam designs. Drop-arm crash beams operate by raising and lowering the beam vertically across the road. Cantilever crash beams are projecting structures that are opened and closed by extending the beam from the hinge buttress to the receiving buttress located on the opposite side of the road. In the swing beam design, the beam is hinged to the buttress such that it swings horizontally across the road. Generally, swing beam and cantilever designs are used at locations where a vertical lift beam is impractical. For example, the swing beam or cantilever designs are utilized at entrances and exits with overhangs, trees, or buildings that would physically block the operation of the drop-arm beam design. Installing any of these crash beam barriers involves the excavation of a pit approximately 48 inches deep for both the hinge and the receiver footings. Due to the depth of excavation, the site should be inspected for underground utilities before digging begins.

In contrast to wedge barriers and crash beams, which are typically installed separately from a fence line, *gates* are often integrated units of a perimeter fence or wall around a facility. Gates are basically movable pieces of fencing that can be opened and closed across a road. When the gate is in the closed (armed) position, the leaves of the gate lock into steel buttresses that are embedded in concrete foundation located on both sides of the roadway, thereby blocking access to the roadway. Generally, gate barricades are constructed from a combination of heavy gauge steel and aluminum that can absorb an impact from vehicles attempting to ram through them. Any remaining impact energy not absorbed by the gate material is transferred to the steel buttresses and their concrete foundation.

Gates can utilize a cantilever, linear, or swing design. Cantilever gates are projecting structures that operate by extending the gate from the hinge footing across the roadway to the receiver footing. A linear gate is designed to slide across the road on tracks via a rack and pinion drive mechanism. Swing gates are hinged so that they can swing horizontally across the road. Installation of the cantilever, linear, or swing gate designs described above involves the excavation of a pit approximately 48 inches deep for both the hinge and receiver footings to which the gates are attached. Due to the depth of excavation, the site should be inspected for underground utilities before digging begins.

Bollards are vertical barriers at least 3 feet tall and 1–2 feet in diameter that are typically set 4–5 feet apart from each other so that they block vehicles from passing between them. Bollards can be fixed in place, removable, or retractable. Fixed and removable bollards are passive barriers that are typically used along building perimeters or on sidewalks to keep vehicles out, while allowing pedestrians to pass.

Plant Security

In contrast to passive bollards, retractable bollards are active security barriers that can easily be raised and lowered to allow vehicles to pass between them. Thus, they can be used in driveways or on roads to control vehicular access. When the bollards are raised, they protect above the road surface and block the roadway; when they are lowered, they sit flush with the road surface, and thus allow traffic to pass over them. Retractable bollards are typically constructed from steel or other materials that have a low weight-to-volume ratio so that they require low power to raise and lower. Steel is also more resistant to breaking than is a more brittle material, such as concrete, and is better able to withstand direct vehicular impact without breaking apart.

Retractable bollards are installed in a trench dug across a roadway—typically at an entrance or gate. Installing retractable bollards requires preparing the road surface. Depending on the vendor, bollards can be installed either in a continuous slab of concrete or in individual excavations with concrete poured in place. The required excavation for a bollard is typically slightly wider and slightly deeper than the bollard height when extended aboveground. The bottom of the excavation is typically lined with gravel to allow drainage. The bollards are then connected to a control panel which controls the raising and lowering of the bollards. Installation typically requires mechanical, electrical, and concrete work; if utility personnel with these skills are available, then the utility can install the bollards themselves.

Portable/removable barriers, which can include removable crash beams and wedge barriers, are mobile obstacles that can be moved in and out of position on a roadway. For example, a crash beam may be completely removed and stored offsite when it is not needed. An additional example would be wedge barriers that are equipped with wheels that can be removed after the barricade is towed into place.

When portable barricades are needed, they can be moved into position rapidly. To provide them with added strength and stability, they are typically anchored to buttress boxes that are located on either side of the road. These buttress boxes, which may or may not be permanent, are usually filled with sand, water, cement, gravel, or concrete to make them heavy and aid in stabilizing the portable barrier. In addition, these buttresses can help dissipate any impact energy from vehicles crashing into the barrier itself.

Because these barriers are not anchored into the roadway, they do not require excavation or other related construction for installation. In contrast, they can be assembled and made operational in a short period of time. The primary shortcoming of this type of design is that these barriers may move if they are hit by vehicles. Therefore, it is important to carefully assess the placement and anchoring of these types of barriers to ensure that they can withstand the types of impacts that may be anticipated at that location.

Because the primary threat to active security barriers is that vehicles will attempt to crash through them, their most important attributes are their size, strength, and crash resistance. Other important features for an active security barrier are the mechanisms by which the barrier is raised and lowered to allow authorized vehicle entry, and other factors, such as weather resistance and safety features.

Passive Security Barriers

One of the most basic threats facing any facility is from intruders accessing the facility with the intention of causing damage to its assets. These threats may include intruders actually entering the facility as well as intruders attacking the facility from outside without actually entering it (i.e., detonating a bomb near enough to the facility to cause damage within its boundaries).

Security barriers are one of the most effective ways to counter the threat of intruders accessing a facility or the facility perimeter. Security barriers are large, heavy structures that are used to control access through a perimeter to either vehicles or personnel. They can be used in many different ways depending on how/where they are located at the facility. For example, security barriers can be used on or along driveways or roads to direct traffic to a checkpoint (i.e., a facility may install jersey barriers in a road to direct traffic in certain direction). Other types of security barriers (crash beams, gates, etc.) can be installed at the checkpoint so that guards can regulate which vehicles can access the facility. Finally, other security barriers (i.e., bollards or security planters) can be used along the facility perimeter to establish a protective buffer area between the facility and approaching vehicles. Establishing such a protective buffer can help in mitigating the effects of the type of bomb blast described above, both by potentially absorbing some of the blast and by increasing the "stand-off" distance between the blast and the facility (the force of an explosion is reduced as the shock wave travels further from the source, and thus the further the explosion is from the target, the less effective it will be in damaging the target).

Security barriers can be either "active" or "passive." "Active" barriers, which include gates, retractable bollards, wedge barriers, and crash barriers, are readily movable, and thus they are typically used in areas where they must be moved often to allow vehicles to pass—such as in roadways at entrances and exits to a facility. In contrast to active security barriers, "passive" security barriers, which include jersey barriers, bollards, and security planters, are not designed to be moved on a regular basis, and thus they are typically used in areas where access is not required or allowed—such as along building perimeters or in traffic control areas. Passive security barriers are typically large, heavy structures that are usually several feet high, and are designed so that even heavy-duty vehicles cannot go over or through them. Therefore, they can be placed in a roadway parallel to the flow of traffic so that they direct traffic in a certain direction (such as to a guardhouse, a gate, or some other sort of checkpoint), or perpendicular to traffic such that they prevent a vehicle from using a road or approaching a building or area.

Biometric Security Systems

Biometrics involves measuring the unique physical characteristics or traits of the human body. Any aspect of the body that is measurably different from person to person—for example, fingerprints or eye characteristics—can serve as a unique biometric identifier for that individual. Biometric systems recognizing fingerprints, palm shape, eyes, face, voice, and signature comprise the bulk of the current biometric systems that recognize other biological features do exist. Biometric security systems use biometric technology combined with some type of locking mechanisms to control access to specific assets. In order to access an asset controlled by a biometric security system, an individual's biometric trait must be matched with an existing profile stored in a database. If there is a match between the two, the locking mechanisms (which could be a physical lock, such as at a doorway, an electronic lock, such as at a computer terminal, or some other type of lock) are disengaged, and the individual is given access to the asset. A biometric security system is typically comprised of the following components:

- A sensor, which measures/records a biometric characteristic or trait
- A control panel, which serves as the connection point between various system components. The control panel communicates information back and forth between the sensor and the host computer, and controls access to the asset by engaging or disengaging the system lock based on internal logic and information from the host computer
- A host computer, which processes and stores the biometric trait in a database
- Specialized software, which compares an individual image taken by the sensor with a stored profile or profiles
- A locking mechanism which is controlled by the biometric system
- A power source to power the system

Biometric Hand and Finger Geometry Recognition

Hand and finger geometry recognition is the process of identifying an individual through the unique "geometry" (shape, thickness, length, width, etc.) of that individual's hand or fingers. Hand geometry recognition has been employed since the early 1980s and is among the most widely used biometric technologies for controlling access to important assets. It is easy to install and use, and is appropriate for use in any location requiring the use of two-finger highly accurate, non-intrusive biometric security. For example, it is currently used in numerous workplaces, day care facilities, hospitals, universities, airports, and power plants.

A newer option within hand geometry recognition technology is finger geometry recognition (not to be confused with fingerprint recognition). Finger geometry recognition relies on the same scanning methods and technologies as does hand geometry recognition, but the scanner only scans two of the user's fingers, as opposed to his entire hand. Finger geometry recognition has been in commercial use since the mid-1990s and is mainly used in time and attendance applications (i.e., to track when individuals have entered and exited a location). To date, the only large-scale commercial use of two-finger geometry for controlling access is at Disney World, where season pass holders use the geometry of their index and middle finger to gain access to the facilities.

Hand and finger geometry recognition systems can be used in several different types of applications, including access control and time and attendance tracking. While time and attendance tracking can be used for security, it is primarily used for operations and payroll purposes (i.e., clocking in and clocking out). In contrast, access control applications are more likely to be security related. Biometric systems are widely used for access control, and can be used on various types of assets, including entryways, computers, vehicles, etc. However, because of their size, hand/finger recognition systems are primarily used in entryway access control applications.

Iris Recognition

The iris, which is the colored or pigmented area of the eye surrounded by the sclera (the white portion of the eye), is a muscular membrane that controls the amount of light entering the eye by contracting or expanding the pupil (the dark center of the eye). The dense, unique patterns of connective tissue in the human iris were first noted in 1936, but it was not unitl1994, when algorithms for iris recognition were created and patented, that commercial applications using biometric iris recognition began to be used extensively. There are now two vendors producing iris recognition technology: both the original developer of these algorithms and a second company which has developed and patented a different set of algorithms for iris recognition.

The iris is an ideal characteristic for identifying individuals because it is formed *in utero*, and its unique patterns stabilize around eight months after birth. No two irises are alike, neither an individual's right or left irises nor the irises of identical twins. The iris is protected by the cornea (the clear covering over the eye), and therefore it is not subject to the aging or physical changes (and potential variation) that are common to some other biometric measures, such as the hand, fingerprints, and the face. Although some limited changes can occur naturally over time, these changes generally occur in the iris's melanin and therefore affect only the eye's color, and not its unique patterns (in addition, because iris scanning uses only black-and-white images, color changes would not affect the scan anyway). Thus, barring specific injuries or certain rare surgeries directly affecting the iris, the iris's unique patterns remain relatively unchanged over an individual's lifetime.

Iris recognition systems employ a monochromatic, or black-and-white, video camera that uses both visible and near infrared light to take video of an individual's iris. Video is used rather than still photography as an extra security procedure. The video is used to confirm the normal continuous fluctuations of the pupil as the eye focuses, which ensures that the scan is of a living human being, and not a photograph or some other attempted hoax. A high-resolution image of the iris is then captured or extracted from the video, using a device often referred to as a *frame grabber*. The unique characteristics identified in this image are then converted into a numeric code, which is stored as a template for that user.

Card Identification/Access/Tracking Systems

A card reader system is a type of electronic identification system that is used to identify a card and then perform an action associated with that card. Depending on the system, the card may identify where a person is or where they were at a certain time, or it may authorize another action, such as disengaging a lock. For example, a security guard may use his card at card readers located throughout a facility to indicate that he has checked a certain location at a certain time. The reader will store the information and/or send it to a central location, where it can be checked later to ensure that the guard has patrolled the area. Other card reader systems can be associated with a lock, so that the cardholder must have their card read and accepted by the reader before the lock disengages. A complete card reader system typically consists of the following components:

- Access cards that are carried by the user
- Card readers, which read the card signals and send the information to control units
- Control units, which control the response of the card reader to the card
- A power source

Plant Security 201

Numerous card reader systems are available. The primary difference between card reader systems lies in the way that data is encoded on the cards and in the way these data are transferred between the card and the card reader, and in the types of applications for which they are best suited. However, all card systems are similar in the way that the card reader and control unit interact to respond to the card.

While card readers are similar in the way that the card reader and control unit interact to control access, they are different in the way data is encoded on the cards and the way these data are transferred between the card and the card reader. There are several types of technologies available for card reader systems. These include:

- Proximity
- Wiegand
- Smartcard
- Magnetic stripe
- Bar code
- Infrared
- Barium ferrite
- Hollerith
- Mixed technologies

The level of security provided by each type (low, moderate, or high) is based on the level of technology a given card reader system has and how simple it is to duplicate that technology, and thus bypass the security. Vulnerability ratings were based on whether the card reader can be damaged easily due to frequent use or difficult working conditions (i.e., weather conditions if the reader is located outside). Often this is influenced by the number of moving parts in the system—the more the moving parts, the greater the system's potential susceptibility to damage. The life-cycle rating is based on the durability of a given card reader system over its entire operational period. Systems requiring frequent physical contact between the reader and the card often have a shorter life cycle due to the wear and tear to which the equipment is exposed. For many card reader systems, the vulnerability and life-cycle ratings have a reciprocal relationship. For instance, if a given system has a high vulnerability rating it will almost always have a shorter life cycle.

Card reader technology can be implemented for facilities of any size and with any number of users. However, because individual systems vary in the complexity of their technology and in the level of security they can provide to a facility, individual users must determine the appropriate system for their needs. Some important features to consider when selecting a card reader system include:

- What level of technological sophistication and security does the card system have?
- How large is the facility, and what are its security needs?
- How frequently will the card system be used? For systems that will experience a high frequency of use it is important to consider a system that has a longer life cycle and lower vulnerability rating, thus making it more cost effective to implement.

- Under what conditions will the system be used? (Will it be installed on the interior or exterior of buildings? Does it require light or humidity controls?) Most card reader systems can operate under normal environmental conditions, and therefore this would be a mitigating factor only in extreme conditions.
- What are the system costs?

Fences

A fence is a physical barrier that can be set up around the perimeter of an asset. Fences often consist of individual pieces (such as individual pickets in a wooden fence, or individual sections of a wrought iron fence) that are fastened together. Individual sections of the fence are fastened together using posts, which are sunk into the ground to provide stability and strength for the sections of the fence hung between them. Gates are installed between individual sections of the fence to allow access inside the fenced area.

Fences are often used as decorative architectural features to separate physical spaces from each other. They may also be used to physically mark the location of a boundary (such as a fence installed along a properly line). However, a fence can also serve as an effective means for physically delaying intruders from gaining access to a water or wastewater asset. For example, many utilities install fences around their primary facilities, around remote pump stations, or around hazardous materials storage areas or sensitive areas within a facility. Access to the area can be controlled through security at gates or doors through the fence (for example, by posting a guard at the gate or by locking it). In order to gain access to the asset, unauthorized persons could have to go either around or through the fence.

Fences are often compared with walls when determining the appropriate system for perimeter security. While both fences and walls can provide adequate perimeter security, fences are often easier and less expensive to install than walls. However, they do not usually provide the same physical strength that walls do. In addition, many types of fences have gaps between the individual pieces that make up the fence (i.e., the spaces between chain links in a chain link fence or the space between pickets in a picket fence). Thus, many types of fences allow the interior of the fenced area to be seen. This may allow intruders to gather important information about the locations or the defenses of vulnerable areas within the facility.

Numerous types of materials are used to construct fences, including chain link iron, aluminum, wood, or wire. Some types of fences, such as split rails or pickets, may not be appropriate for security purposes because they are traditionally low fences, and they are not physically strong. Potential intruders may be able to easily trespass these fences either by jumping or by climbing over them or by breaking through them. For example, the rails in a split fence may be broken easily.

Important security attributes of a fence include the height to which it can be constructed, the strength of the material comprising the fence, the method and strength of attaching the individual sections of the fence together at the posts and the fence's ability to restrict the view of the assets inside the fence. Additional considerations should include the ease of installing the fence and the ease of removing and reusing sections of the fence.

Plant Security

Some fences can include additional measures to delay, or even detect, potential intruders. Such measures may include the addition of barbed wire, razor wire, or other deterrents at the top of the fence. Barbed wire is sometimes employed at the base of fences as well. This can impede a would-be intruder's progress in even reaching the fence. Fences may also be fitted with security cameras to provide visual surveillance of the perimeter. Finally, some facilities have installed motion sensors along their fences to detect movement on the fence. Several manufacturers have combined these multiple perimeter security features into one product and offer alarms and other security features.

The correct implementation of a fence can make it a much more effective security measure. Security experts recommend the following when a facility constructs a fence:

- The fence should be at least 7–9 feet high.
- Any outriggers, such as barbed wire, that are affixed on top of the fence should be angled out and away from the facility, and not in toward the facility. This will make climbing the fence more difficult, and will prevent ladders from being placed against the fence.
- Other types of hardware can increase the security of the fence. This can include installing concertina wire along the fence (this can be done in front of the fence or at the top of the fence), or adding intrusion sensors, camera, or other hardware to the fence.
- All undergrowth should be cleared for several feet (typically 6 feet) on both sides of the fence. This will allow for a clearer view of the fence by any patrols in the area.
- Any trees with limbs or branches hanging over the fence should be trimmed so that intruders cannot use them to go over the fence. Also, it should be noted that fallen trees can damage fences, and so management of trees around the fence can be important. This can be especially important in areas where the fence goes through a remote area.
- Fences that do not block the view from outside allow patrols to see inside the fence without having to enter the facility.
- "No Trespassing" signs posted along the fence can be a valuable tool in prosecuting any intruders who claim that the fence was broken and that they did not enter through the fence illegally. Adding signs that highlight the local ordinances against trespassing can further dissuade simple troublemakers from illegally jumping/climbing the fence.

Films for Glass Shatter Protection

Most water and wastewater utilities have numerous windows on the outside of buildings, on doors, and in interior offices. In addition, many facilities have glass doors or other glass structures, such as glass walls or display cases. These glass objects are potentially vulnerable to shattering when heavy objects are thrown or launched at them, when explosions occur near them, or when there are high winds (for exterior glass). If the glass is shattered, intruders may potentially enter an area. In addition, shattered glass projected into a room from an explosion or from an object being thrown through a door or window can injure and potentially incapacitate personnel in the room. Materials that prevent glass from shattering can help to maintain the integrity of the door, window, or other glass objects, and can delay an intruder from gaining access. These materials can also prevent flying glass and thus reduce potential injuries.

Materials designed to prevent glass from shattering include specialized films and coatings. These materials can be applied to existing glass objects to improve their strength and their ability to resist shattering. The films have been tested against many scenarios that could result in glass breakage, including penetration by blunt objects, bullets, high winds, and simulated explosions. Thus, the films are tested against both simulated weather scenarios (which could include both the high winds themselves and the force of objects blown onto the glass) and more criminal/terrorist scenarios where the glass is subject to explosives or bullets. Many vendors provide information on the results of these types of tests, and thus potential users can compare different product lines to determine which products best suit their needs.

The primary attributes of films used in shatter protection are:

- The materials from which the film is made
- The adhesive that bonds the film to the glass surface
- The thickness of the film

Fire Hydrant Locks

Fire hydrants are installed at strategic locations throughout a community's water distribution system to supply water for firefighting. However, because there are many hydrants in a system and they are often located in residential neighborhood, industrial districts, and other areas where they cannot be easily observed and/or guarded, they are potentially vulnerable to unauthorized access. Many municipalities, states, and EPA Regions have recognized this potential vulnerability and have instituted programs to lock hydrants. For example, EPA Region 1 has included locking hydrants as number 7 on its "Drinking Water Security and Emergency Preparedness" Top Ten List for small groundwater suppliers.

A "hydrant lock" is a physical security device designed to prevent unauthorized access to the water supply through a hydrant. They can also ensure water and water pressure availability to firefighters and prevent water theft and associated lost water revenue. These locks have been successfully used in numerous municipalities and in various climates and weather conditions.

Fire hydrant locks are basically steel covers or caps that are locked in place over the operating nut of a fire hydrant. The lock prevents unauthorized persons from accessing the operating nut and opening the fire hydrant valve. The lock also makes it more difficult to remove the bolts from the hydrant and access the system that way. Finally, hydrant locks shield the valve from being broken off. Should a vandal attempt to breach the hydrant lock by force and succeed in breaking the hydrant lock, the vandal will only succeed in bending the operating valve. If the hydrant's operating valve is bent, the hydrant will not be operational, but the water asset remains protected and inaccessible to vandals. However, the entire hydrant will need to be replaced.

Hydrant locks are designed so that the hydrants can be operated by special "key wrenches" without removing the lock. These specialized wrenches are generally distributed to the fire department, public works department, and other authorized persons so that they can access the hydrants as needed. An inventory of wrenches and their serial numbers is generally kept by a municipality so that the location of all wrenches is known. These operating key wrenches may only be purchased by registered lock owners.

The most important features of hydrants are their strength and the security of their locking systems. The locks must be strong so that they cannot be broken off. Hydrant locks are constructed from stainless or alloyed steel. Stainless steel locks are stronger and are ideal for all climates; however, they are more expensive than alloy locks. The locking mechanism for each fire hydrant locking system ensures that the hydrant can only be operated by authorized personnel who have the specialized key to work the hydrant.

Hatch Security

A hatch is basically a door installed on a horizontal plane (such as in a floor, a paved lot, or a ceiling), instead of on a vertical plane (such as in a building wall). Hatches are usually used to provide access to assets that are either located underground (such as hatches to basements or underground storage areas) or located above ceilings (such as emergency roof exits). At water and wastewater facilities, hatches are typically used to provide access to underground vaults containing pumps, valves, or piping, or to the interior of water tanks or covered reservoirs. Securing a hatch by locking it or upgrading materials to give the hatch added strength can help to delay unauthorized access to any asset behind the hatch. Like all doors, a hatch consists of a frame anchored to the horizontal structure, a door or doors, hinges connecting the door/doors to the frame, and a latching or locking mechanism that keeps the hatch door/doors closed.

It should be noted that improving hatch security is straightforward, and that hatches with upgraded security features can be installed new or they can be retrofit for existing applications. Many municipalities already have specifications for hatch security at their water and wastewater utility assets.

Depending on the application, the primary security-related attributes of a hatch are the strength of the door and frame, its resistance to the elements and corrosion, it ability to be sealed against water or gas, and its locking features. Hatches must be both strong and lightweight so that they can withstand typical static loads (such as people or vehicles walking or driving over them) while still being easy to open. In addition, because hatches are typically installed at outdoor locations, they are usually designed from corrosion-resistant metal that can withstand the elements. Therefore, hatches are typically constructed from high gauge steel or lightweight aluminum.

The hatch locking mechanism is perhaps the most important part of hatch security. There are a number of locks that can be implemented for hatches, including:

- Slam locks (internal locks that are located within the hatch frame)
- Recessed cylinder locks
- Bolt locks
- Padlocks

Intrusion Sensors

An exterior intrusion sensor is a detection device that is used in an outdoor environment to detect intrusions into a protected area. These devices are designed to detect an intruder, and then communicate an alarm signal to an alarm system. The alarm system can respond to the intrusion in many different ways, such as by triggering an audible or visual alarm signal, or by sending an electronic signal to a central monitoring location that notifies security personnel of the intrusion. Intrusion sensors can be used to protect many kinds of assets. Intrusion sensors that protect physical space are classified according to whether they protect indoor, or "interior," spaces (i.e., an entire building or room within a building), or outdoor, or "exterior," spaces (i.e., a fence line or perimeter). Interior intrusion sensors are designed to protect the interior space of a facility by detecting an intruder who is attempting to enter, or who has already entered a room or building. In contrast, exterior intrusion sensors are designed to detect an intrusion into a protected outdoor/exterior area. Exterior protected areas are typically arranged as zones or exclusion areas placed so that the intruder is detected early in the intrusion attempt before they can gain access to more valuable assets (e.g., into a building located within the protected area). Early detection creates additional time for security forces to respond to the alarm.

Buried Exterior Intrusion Sensors

Buried sensors are electronic devices that are designed to detect potential intruders. The sensors are buried along the perimeters of sensitive assets and are able to detect intruder activity both above- and belowground. Some of these systems are composed of individual, stand-alone sensor units, while other sensors consist of buried cables.

Ladder Access Control

Water and wastewater utilities have a number of assets that are raised above ground level, including raised water tanks, raised chemical tanks, raised piping systems, and roof access points into buildings. In addition, communications equipment, antennae, or other electronic devices may be located on the top of these raised assets. Typically, these assets are reached by ladders that are permanently anchored to the asset. For example, raised water tanks typically are accessed by ladders that are bolted to one of the legs of the tank. Controlling access to these raised assets by controlling access to the ladder can increase security at a water or wastewater utility.

A typical ladder access control system consists of some type of cover that is locked or secured over the ladder. The cover can be a casing that surrounds most of the ladder, or a door or shield that covers only part of the ladder. In either case, several rungs of the ladder (the number of rungs depends on the size of the cover) are made inaccessible by the cover, and these rungs can only be accessed by opening or removing the cover. The cover is locked so that only authorized personnel can open or remove it and use the ladder. Ladder access controls are usually installed at several feet above ground level, and they usually extend several feet up the ladder so that they cannot be circumvented by someone accessing the ladder above the control system. The important features of ladder access control are the size and strength of the cover and its ability to lock or otherwise be secured from unauthorized access.

The covers are constructed from aluminum or some type of steel. This should provide adequate protection from being pierced or cut through. The metals are corrosion resistant so that they will not corrode or become fragile from extreme weather conditions in outdoor applications. The bolts used to install each of these systems are galvanized steel. In addition, the bolts for each cover are installed on the inside of the unit so they cannot be removed from the outside.

Locks

A lock is a type of physical security device that can be used to delay or prevent a door, a window, a manhole, a filing cabinet drawer, or some other physical feature from being opened, moved, or operated. Locks typically operate by connecting two pieces together—such as by connecting a door to a door jamb or a manhole to its casement. Every lock has two modes—engaged (or "locked") and disengaged (or "opened"). When a lock is disengaged, the asset on which the lock is installed can be accessed by anyone, but when the lock is engaged, only authorized personnel can access the locked asset.

Locks are excellent security features because they have been designed to function in many ways and to work on many different types of assets. Locks can also provide different levels of security depending on how they are designed and implemented. The security provided by a lock is dependent on several factors, including its ability to withstand physical damage (i.e., can it be cut off, broken, or otherwise physically disabled) as well as its requirements for supervision or operation (i.e., combinations may need to be changed frequently so that they are not compromised and the locks remain secure). While there is no single definition of the "security" of a lock, locks are often described as minimum, medium, or maximum security. Minimum security locks are those that can be easily disengaged (or "picked") without the correct key or code, or those that can be disabled easily (such as small padlocks that can be cut with bolt cutters). Higher security locks are more complex and thus are more difficult to pick, or are sturdier and more resistant to physical damage.

Many locks, such as many door locks, only need to be unlocked from one side. For example, most door locks need a key to be unlocked only from the outside. A person opens such devices, called single-cylinder locks, from the inside by pushing a button or by turning a knob or handle. Double-cylinder locks require a key to be locked or unlocked from both sides.

Manhole Intrusion Sensors

Manholes are located at strategic locations throughout most municipal water, wastewater, and other underground utility systems. Manholes are designed to provide access to the underground utilities, and therefore they are potential entry points to a system. For example, manholes in water or wastewater systems may provide access to sewer lines or vaults containing on/off or pressure reducing water valves. Because many utilities run under other infrastructure (roads, building), manholes also provide potential access points to critical infrastructure as well as water and wastewater assets. In addition, because the portion of the system to which manholes provide entry is primarily located underground, access to a system through a manhole increases the chance that an intruder will not be seen.

Therefore, protecting manholes can be a critical component of guarding an entire community.

The various methods for protecting manholes are designed to prevent unauthorized personnel from physically accessing the manhole, and detecting attempts at unauthorized access to the manhole. A manhole intrusion sensor is a physical security device designed to detect unauthorized access to the utility through a manhole. Monitoring a manhole that provides access to a water or wastewater system can mitigate two distinct types of threats. First, monitoring a manhole may detect access of unauthorized personnel to water or wastewater systems or assets through the manhole. Second, monitoring manholes may also allow the detection of the introduction of hazardous substances into the water system.

Several different technologies have been used to develop manhole intrusion sensors, including mechanical systems, magnetic systems, and fiber optic and infrared sensors. Some of these intrusion sensors have been specifically designed for manholes, while others consist of standard, off-the-shelf intrusion sensors that have been implemented in a system specifically designed for application in a manhole.

Manhole Locks

A manhole lock is a physical security device designed to delay unauthorized access to the utility through a manhole. Locking a manhole that provides access to a water or wastewater system can mitigate two distinct types of threats. First, locking a manhole may delay unauthorized personnel from accessing water or wastewater systems through the manhole. Second, locking manholes may also prevent the introduction of hazardous substances into the wastewater or stormwater system.

Radiation Detection Equipment for Monitoring Personnel and Packages

A major potential threat facing water and wastewater facilities is contamination by radioactive substances. Radioactive substances brought on-site at a facility could be used to contaminate the facility, thereby preventing workers from safely entering the facility to perform necessary water treatment tasks. In addition, radioactive substances brought on-site at a water treatment plant could be discharged into the water source or the distribution system, contaminating the downstream water supply. Therefore, detection of radioactive substances being brought on-site can be an important security enhancement.

Various radionuclides have unique properties, and different equipment is required to detect different types of radiation. However, it is impractical and potentially unnecessary to monitor for specific radionuclides being brought on-site. Instead, for security purposes, it may be more useful to monitor for gross radiation as an indicator of unsafe substances.

In order to protect against these radioactive materials being brought on-site, a facility may set up monitoring sites outfitted with radiation detection instrumentation at entrances to the facility. Depending on the specific types of equipment chosen, this equipment would detect radiation emitted from people, packages, or other objects being brought through an entrance.

One of the primary differences between the different types of detection equipment is the means by which the equipment reads the radiation. Radiation may be detected either by direct measurement or through sampling. Direct radiation measurement involves measuring radiation through an external probe on the detection instrumentation. Some direct measurement equipment detects radiation emitted into the air around the monitored object. Because this equipment detects radiation in the air, it does not require that the monitoring equipment make physical contact with the monitored object. Direct means for detecting radiation include using a walk-through portal-type monitor that would detect elevated radiation levels on a person or in a package, or by using a handheld detector, which would be moved or swept over individual objects to locate a radioactive source.

Some types of radiation, such as alpha or low energy beta radiation, have a short range and are easily shielded by various materials. These types of radiation cannot be measured through direct measurement. Instead, they must be measured through sampling. Sampling involves wiping the surface to be tested with a special filter cloth, and then reading the cloth in a special counter. For example, specialized smear counters measure alpha and low energy beta radiation.

Reservoir Covers

Reservoirs are used to store raw or untreated water. They can be located underground (buried), at ground level, or on an elevated surface. Reservoirs can vary significantly in size; small reservoirs can hold as little as 1,000 gallons, while larger reservoirs may hold many million gallons. Reservoirs can be either natural or man-made. Natural reservoirs can include lakes or other contained water bodies, while man-made reservoirs usually consist of some sort of engineered structure, such as a tank or other impoundment structure. In addition to the water containment structure itself, reservoir systems may also include associated water treatment and distribution equipment, including intakes, pumps, pump houses, piping systems, chemical treatment, and chemical storage areas.

Drinking water reservoirs are of particular concern because they are potentially vulnerable to contamination of the stored water, either through direct contamination of the storage area or through infiltration of the equipment, piping, or chemicals associated with the reservoir. For example, because many drinking water reservoirs are designed as aboveground, open-air structures, they are potentially vulnerable to airborne deposition, bird and animal wastes, human activities, and dissipation of chlorine or other treatment chemicals. However, one of the most serious potential threats to the system is direct contamination of the stored water through dumping contaminants into the reservoir. Utilities have taken various measures to mitigate this type of threat, including fencing off the reservoir, installing cameras to monitor for intruders, and monitoring for changes in water quality. Another option for enhancing security is covering the reservoir using some type of manufactured cover to prevent intruders from gaining physical access to the stored water. Implementing a reservoir cover may or may not be practical depending on the size of the reservoir (for example, covers are not typically used on natural reservoirs because they are too large for the cover to be technically feasible and cost effective). This section will focus on drinking water reservoir covers, where and how they are typically implemented, and how they can be used to reduce the threat of contamination of the stored water. While covers can enhance the reservoir's security, it should be noted that covering a reservoir typically changes the reservoir's operational requirements. For example, vents must be installed in the cover to ensure gas exchange between the stored water and the atmosphere.

A reservoir cover is a structure installed on or over the surface of the reservoir to minimize water quality degradation. The three basic design types for reservoir covers are:

- Floating
- Fixed
- Air-supported

A variety of materials are used when manufacturing a cover, including reinforced concrete, steel, aluminum, polypropylene, chlorosulfonated polyethylene, or ethylene interpolymer alloys. There are several factors that affect a reservoir cover's effectiveness, and thus its ability to protect the stored water. These factors include:

- The location, size, and shape of the reservoir
- The ability to lay/support a foundation (for example, footing, soil, and geotechnical support conditions)
- The length of time the reservoir can be removed from service for cover installation or maintenance
- Aesthetic considerations
- Economic factors, such as capital and maintenance costs

It may not be practical, for example, to install a fixed cover over a reservoir if the reservoir is too large or if the local soil conditions cannot support a foundation. A floating or air-supported cover may be more appropriate for these types of applications.

In addition to the practical considerations for installation of these types of covers, there are a number of operations and maintenance (O&M) concerns that affect the utility of a cover for specific applications, including how different cover materials will withstand local climatic conditions, what types of cleaning and maintenance will be required for each particular type of cover, and how these factors will affect the cover's lifespan and its ability to be repaired when it is damaged.

The primary feature affecting the security of a reservoir cover is its ability to maintain its integrity. Any type of cover, no matter what its construction material, will provide good protection from contamination by rainwater or atmospheric deposition, as well as from intruders attempting to access the stored water with the intent of causing intentional contamination. The covers are large and heavy, and it is difficult to circumvent them to get into the reservoir. At the very least, it would take a determined intruder, as opposed to a vandal, to surpass the cover.

Side-Hinged Door Security

Doorways are the main access points to a facility or to rooms within a building. They are used on the exterior or in the interior of buildings to provide privacy and security for the areas behind them. Different types of doorway security systems may be installed in different doorways depending on the needs or requirements of the buildings or rooms. For example, exterior doorways tend to have heavier doors to withstand the elements and to provide some security to the entrance of the building. Interior doorways in office areas may have lighter doors that may be primarily designed to provide privacy rather than security. Therefore, these doors may be made of glass or lightweight wood. Doorways in industrial areas may have sturdier doors than do other interior doorways and may be designed to provide protection or security for areas behind the doorway. For example, fireproof doors may be installed in chemical storage areas or in other areas where there is a danger of fire. Because they are the main entries into a facility or a room, doorways are often prime targets for unauthorized entry into a facility or an asset. Therefore, securing doorways may be a major step in providing security at a facility. A doorway includes four main components:

- The door, which blocks the entrance. The primary threat to the actual door is breaking or piercing through the door. Therefore, the primary security features of doors are their strength and resistance to various physical threats, such as fire or explosions.
- The door frame, which connects the door to the wall. The primary threat to a door frame is that the door can be pried away from the frame. Therefore, the primary security feature of a door frame is its resistance to prying.
- The hinges, which connect the door to the door frame. The primary threat to door hinges is that they can be removed or broken, which will allow intruders to remove the entire door. Therefore, security hinges are designed to be resistant to breaking. They may also be designed to minimize the threat of removal from the door.
- The lock, which connects the door to the door frame. Use of the lock is controlled through various security features, such as keys, combinations, etc., such that only authorized personnel can open the lock and go through the door. Locks may also incorporate other security features, such as software or other systems to track overall use of the door or to track individuals using the door, etc.

Each of these components is integral to providing security for a doorway, and upgrading the security of only one of these components while leaving the other components remain unprotected may not increase the overall security of the doorway. For example, many facilities upgrade door locks as a basic step in increasing the security of a facility. However, if the facilities do not also focus on increasing security for the door hinges or the door frame, the door may remain vulnerable to being removed from its frame, thereby defeating the purpose of the increased security of the door lock.

The primary attribute for the security of a door is its strength. Many security doors are 4–20 gauge hollow metal doors consisting of steel plates over a hollow cavity reinforced with steel stiffeners to give the doors extra stiffness and rigidity. This increases resistance to blunt force used to penetrate through the door. The space between the stiffeners may be filled with specialized materials to provide the door resistance against fire, blast, or bullet. The Window and Door Manufacturers

Association has developed a series of performance attributes for doors. These include:

- Structural resistance
- Forced entry resistance
- Hinge style screw resistance
- Split resistance
- Hinge resistance
- Security rating
- Fire resistance
- Bullet resistance
- Blast resistance

The first five attributes provide information on a door's resistance to standard physical breaking and prying attacks. These tests are used to evaluate the strength of the door and the resistance of the hinges and the frame in a standardized way. For example, the rack load test simulates a prying attack on a corner of the door. A test panel is restrained at one end, and a third corner is supported. Loads are applied and measured at the fourth corner. The door impact test simulates a battering attack on a door and frame using impacts of 200 foot pounds by a steel pendulum. The door must remain fully operable after the test. It should be noted that door glazing is also rated for resistance to shattering, etc. Manufacturers will be able to provide security ratings for these features of a door as well.

Door frames are an integral part of doorway security because they anchor the door to the wall. Door frames are typically constructed from wood or steel, and they are installed such that they extend several inches over the doorway that has been cut into the wall. For added security, frames can be designed to have varying degrees of overlap with, or wrapping over, the underlying wall. This can make prying the frame from the wall more difficult. A frame formed from a continuous piece of metal (as opposed to a frame constructed from individual metal pieces) will prevent prying between pieces of the frame.

Many security doors can be retrofit into existing frames; however, many security door installations include replacing the door frame or the door itself. For example, bullet resistance per Underwriter's Laboratory 752 requires resistance of the door and frame assembly, and thus replacing the door only would not meet UL 752 requirements.

Valve Lockout Devices

Valves are utilized as control elements in water and wastewater process piping networks. They regulate the flow of both liquids and gases by opening, closing, or obstructing a flow passageway. Valves are typically located where flow control is necessary. They can be located in-line or at pipeline and tank entrance and exit points. They can serve multiple purposes in a process pipe network, including:

- Redirecting and throttling flow
- Preventing backflow

Plant Security

- Shutting off flow to a pipeline or tank (for isolation purposes)
- Releasing pressure
- Draining extraneous liquid from pipelines or tanks
- Introducing chemicals into the process network
- Acting as access points for sampling process water

Valves are located at critical junctures throughout water and wastewater systems, both on-site at treatment facilities and off-site within water distribution and wastewater collection systems. They may be located either aboveground or belowground. Because many valves are located within the community, it is critical to provide protection against valve tampering. For example, tampering with a pressure relief valve could result in a pressure buildup and potential explosion in the piping network. On a larger scale, addition of a pathogen or chemical to the water distribution system through an unprotected valve could result in the release of that contaminant to the general population.

Various security products are available to protect aboveground and belowground valves. For example, valve lockout devices can be purchased to protect valves and valve controls located aboveground. Vaults containing underground valves can be locked to prevent access to these valves. Valve-specific lockout devices are available in a variety of colors, which can be useful in distinguishing different valves. For example, different-colored lockouts can be used to distinguish the type of liquid passing through the valve (i.e., treated, untreated, potable, chemical), or to identify the party responsible for maintaining the lockout. Implementing a system of different-colored locks on operating valves can increase system security by reducing the likelihood of an operator inadvertently opening the wrong valve and causing a problem in the system.

Vent Security

Vents are installed in aboveground, covered water reservoirs and in underground reservoirs to allow ventilation of the stored water. Specifically, vents permit the passage of air that is being displaced from, or drawn into, the reservoir as the water level in the reservoir rises and falls due to system demands. Small reservoirs may require only one vent, whereas larger reservoirs may have multiple vents throughout the system.

The specific vent design for any given application will vary depending on the design of the reservoir, but every vent consists of an open-air connection between the reservoir and the outside environment. Although these air-exchange vents are an integral part of covered or underground reservoirs, they also represent a potential security threat. Improving vent security by making the vents tamper resistant or by adding other security features, such as security screens or security covers, can enhance the security of the entire water system. Many municipalities already have specifications for vent security at their water assets. These specifications typically include the following requirements:

- Vent openings are to be angled down or shielded to minimize the entrance of surface and/or rainwater into the vent through the opening.

- Vent designs are to include features to exclude insects, birds, animals, and dust.
- Corrosion-resistant materials are to be used to construct the vents.

Some states have adopted more specific requirements for added vent security at their water utility assets. For example, the State of Utah's Department of Environmental Quality, Division of Drinking Water, Division of Administrative Rules (DAR), provides specific requirements for public drinking water storage tanks. The rules for drinking water storage tanks as they apply to venting are set forth in Utah-R309-545-15: "Venting," and include the following requirements:

- Drinking water storage tank vents must have an open discharge on buried structures
- The vents must be located 24–36 inches above the earthen covering
- The vents must be located and sized to avoid blockage during winter conditions

In a second example, Washington State's "Drinking Water Tech Tips: Sanitary Protection of Reservoirs" document states that vents must be protected to prevent the water supply from being contaminated. The document indicates that non-corrodible No. 4 mesh may be used to screen vents on elevated tanks. The document continues to state that the vent opening for storage facilities located underground or at ground level should be 24–36 inches above the roof or ground and that it must be protected with a No. 24-inch mesh non-corrodible screen. The New Mexico Administrative Code also specifies that vents must be covered with No. 24 mesh (NMAC Title 20, Chapter 7, Subpart I, 208.E). Washington and New Mexico, as well as many other municipalities, require vents to be screened using a non-corrodible mesh to minimize the entry of insects, other animals, and rain-borne contamination into the vents. When selecting the appropriate mesh size, it is important to identify the smallest mesh size that meets both the strength and durability requirements for that application.

Visual Surveillance Monitoring

Visual surveillance is used to detect threats through continuous observation of important or vulnerable areas of an asset. The observations can also be recorded for later review or use (for example, in court proceedings). Visual surveillance system can be used to monitor various parts of collection, distribution, or treatment systems, including the perimeter of a facility, outlying pumping stations, or entry or access points into specific buildings. These systems are also useful in recording individuals who enter or leave a facility, thereby helping to identify unauthorized access. Images can be transmitted live to a monitoring station, where they can be monitored in real time, or recorded and reviewed later. Many facilities have found that a combination of electronic surveillance and security guards provides an effective means of facility security. Visual surveillance is provided through a closed circuit television (CCTV) system, in which the capture, transmission, and reception of an image is localized within a closed "circuit." This is different than other broadcast images,

Plant Security

such as over-the-air television, which is broadcast over the air to any receiver within range. At a minimum, a CCTV system consists of:

- One or more cameras
- A monitor for viewing the images
- A system for transmitting the images from the camera to the monitor.

WATER MONITORING DEVICES[*]

Earlier it was pointed out that proper security preparation really comes down to a three-legged approach: Detect, Delay, Respond. The third leg of security, to detect, is discussed in this section. Specifically, this section deals with the monitoring of water samples to detect toxicity and/or contamination. Many of the major monitoring tools that can be used to identify anomalies in process streams or finished water that may represent potential threats are discussed, including:

- Sensors for monitoring chemical, biological, and radiological contamination
- Chemical sensor—arsenic measurement system
- Chemical sensor for toxicity (adapted biochemical oxygen demand (BOD) analyzer)
- Chemical sensor—total organic carbon analyzer
- Chemical sensor—chlorine measurement system
- Chemical sensor—portable cyanide analyzer
- Portable field monitors to measure VOCs
- Radiation detection equipment
- Radiation detection equipment for monitoring water assets
- Toxicity monitoring/toxicity meters

Water quality monitoring sensor equipment may be used to monitor key elements of water or wastewater treatment processes (such as influent water quality, treatment processes, or effluent water quality) to identify anomalies that may indicate threats to the system. Some sensors, such as sensors for biological organisms or radiological contaminants, measure potential contamination directly, while others, particularly some chemical monitoring systems, measure "surrogate" parameters that may indicate problems in the system but do not identify sources of contamination directly. In addition, sensors can provide more accurate control of critical components in water and wastewater systems and may provide a means of early warning so that the potential effects of certain types of attacks can be mitigated. One advantage of using chemical and biological sensors to monitor for potential threats to water and wastewater systems is that many utilities already employ sensors to monitor potable water (raw or finished) or influent/effluent for Safe Drinking Water Act (SDWA) or Clean Water Act (CWA) water quality compliance or process control.

---

[*] The following information is adapted from Spellman, R.R., *Water Infrastructure Protection and Homeland Security*, Government Institutes Press, Lanham, Md, 2007.

Chemical sensors that can be used to identify potential threats to water and wastewater systems include inorganic monitors (e.g., chlorine analyzer), organic monitors (e.g., total organic carbon analyzer), and toxicity meters. Radiological meters can be used to measure concentrations of several different radioactive species. Monitors that use biological species can be used as sentinels for the presence of contaminants of concern, such as toxics. At the present time, biological monitors are not in widespread use and very few bio-monitors are used by drinking water utilities in the United States.

Monitoring can be conducted using either portable or fixed-location sensors. Fixed-location sensors are usually used as part of a continuous, on-line monitoring system. Continuous monitoring has the advantage of enabling immediate notification when there is an upset. However, the sampling points are fixed and only certain points in the system can be monitored. In addition, the number of monitoring locations needed to capture the physical, chemical, and biological complexity of a system can be prohibitive. The use of portable sensors can overcome this problem of monitoring many points in the system. Portable sensors can be used to analyze grab samples at any point in the system but have the disadvantage that they provide measurements only at one point in time.

Sensors for Monitoring Chemical, Biological, and Radiological Contamination

Toxicity tests measure water toxicity by monitoring adverse biological effects on test organisms. Toxicity tests have traditionally been used to monitor wastewater effluent streams for National Pollutant Discharge Elimination System (NPDES) permit compliance or to test water samples for toxicity. However, this technology can also be used to monitor drinking water distribution systems or other water/wastewater streams for toxicity. Currently, several types of bio-sensors and toxicity tests are being adapted for use in the water/wastewater security field. The keys to using bio-monitoring or bio-sensors for drinking water or other water/wastewater asset security are rapid response and the ability to use the monitor at critical locations in the system, such as in water distribution systems downstream of pump stations, or prior to the biological process in a wastewater treatment plant. While there are several different organisms that can be used to monitor for toxicity (including bacteria, invertebrates, and fish), bacteria-based bio-sensors are ideal for use as early warning screening tools for drinking water security because bacteria usually respond to toxics in a matter of minutes. In contrast to methods using bacteria, toxicity screening methods that use higher-level organisms such as fish may take several days to produce a measurable result. Bacteria-based bio-sensors have recently been incorporated into portable instruments, making rapid response and field-testing practical. These portable meters detect decreases in biological activity (e.g., decreases in bacterial luminescence), which highly correlate with increased levels of toxicity.

At the present time, few utilities are using biologically based toxicity monitors to monitor water/wastewater assets for toxicity, and very few products are now commercially available. Several new approaches to the rapid monitoring of microorganisms for security purposes (e.g., microbial source tracking) have been identified. However, most of these methods are still in the research and development phase.

Chemical Sensors: Arsenic Measurement System

Arsenic is an inorganic toxin that occurs naturally in soils. It can enter water supplies from many sources, including erosion of natural deposits; runoff from orchards, runoff from glass and electronics production wastes; or leaching from products treated with arsenic, such as wood. Synthetic organic arsenic is also used in fertilizer. Arsenic toxicity primarily associated with inorganic arsenic ingestion has been linked to cancerous health effects, including cancer of the bladder, lungs, skin, kidney, nasal passages, liver, and prostate. Arsenic ingestion has also been linked to noncancerous cardiovascular, pulmonary, immunological, and neurological, endocrine problems. According to the USEPA's Safe Drinking Water Act Arsenic Rule, inorganic arsenic can exert toxic effects after acute (short-term) or chronic (long-term) exposure. Toxicological data for acute exposure, which is typically given as a LD50 value (the dose that would be lethal to 50 percent of the test subjects in a given test), suggests that the LD50 of arsenic ranges 1–4 milligrams of arsenic per kilogram (mg/kg) of body weight. This dose would correspond to a lethal dose range of 70–280 mg for 50 percent of adults weighing 70 kg. At nonlethal, but high, acute doses, inorganic arsenic can cause gastroenterological effects, shock, neuritis (continuous pain), and vascular effects in humans. The USEPA has set a maximum contaminant level goal of 0 for arsenic in drinking water; the current enforceable maximum contaminant level is 0.050 mg/L. As of January 23, 2006, the enforceable MCL for arsenic is 0.010 mg/L.

The SDWA requires arsenic monitoring for public water systems. The Arsenic Rule indicates that surface water systems must collect one sample annually and groundwater systems must collect one sample in each compliance period (once every three years). Samples are collected at entry points to the distribution system, and analysis is done in the lab using one of several USEPA-approved methods, including inductively coupled plasma mass spectroscopy (ICP-MS, USEPA 200.8) and several atomic absorption (AA) methods. However, several different technologies, including colorimetric test kits and portable chemical sensors, are currently available for monitoring inorganic arsenic concentrations in the field. These technologies can provide a quick estimate of arsenic concentrations in a water sample. Thus, these technologies may be useful for spot-checking different parts of a drinking water system (for example, reservoirs and isolated areas of distribution systems) to ensure that the water is not contaminated with arsenic.

Chemical Sensors: Adapted BOD Analyzer

One manufacturer has adapted a BOD analyzer to measure oxygen consumption as a surrogate for general toxicity. The critical element in the analyzer is the bioreactor, which is used to continuously measure the respiration of the biomass under stable conditions. As the toxicity of the sample increases, the oxygen consumption in the sample decreases. An alarm can be programmed to sound if oxygen reaches a minimum concentration (i.e., if the sample is strongly toxic). The operator must then interpret the results into a measure of toxicity. Note that, at the current time, it is difficult to directly define the sensitivity and/or the detection limit of toxicity measurement devices because limited data is available regarding specific correlation of decreased oxygen consumption and increased toxicity of the sample.

Chemical Sensors: Total Organic Carbon Analyzer

Total organic carbon (TOC) analysis is a well-defined and commonly used methodology that measures the carbon content of dissolved and particulate organic matter present in water. Many water utilities monitor TOC to determine raw water quality or to evaluate the effectiveness of processes designed to remove organic carbon. Some wastewater utilities also employ TOC analysis to monitor the efficiency of the treatment process. In addition to these uses for TOC monitoring, measuring changes in TOC concentrations can be an effective "surrogate" for detecting contamination from organic compounds (e.g., petrochemicals, solvents, and pesticides). Thus, while TOC analysis does not give specific information about the nature of the threat, identifying changes in TOC can be a good indicator of potential threats to a system. TOC analysis includes inorganic carbon removal by oxidation of the organic carbon into CO_2, and quantification of the CO_2. The primary differences between different on-line TOC analyzers are in the methods used for oxidation and CO_2 quantification.

The oxidation step can be performed at a high or low temperature. The determination of the appropriate analytical method (and thus the appropriate analyzer) is based on the expected characteristics of the wastewater sample (TOC concentrations and the individual components making up the TOC fraction). In general, high temperature (combustion) analyzers achieve more complete oxidation of the carbon fraction than do low temperature (wet chemistry/UV) analyzers. This can be important both in distinguishing different fractions of the organics in a sample and in achieving a precise measurement of the organic content of the sample. Three different methods are also available for detection and quantification of carbon dioxide produced in the oxidation step of a TOC analyzer. These are:

- Nondispersive infrared (NDIR) detector
- Colorimetric methods
- Aqueous conductivity methods

The most common detector that on-line TOC analyzers use for source water and drinking water analysis is the nondispersive infrared detector.

Although the differences in analytical methods employed by different TOC analyzers may be important in compliance or process monitoring, high levels of precision and the ability to distinguish specific organic fractions from a sample may not be required for detection of a potential chemical threat. Instead, gross deviations from normal TOC concentrations may be the best indication of a chemical threat to the system.

The detection limit for organic carbon depends on the measurement technique used (high or low temperature) and the type of analyzer. Because TOC concentrations are simply surrogates that can indicate potential problems in a system, gross changes in these concentrations are the best indicators of potential threats. Therefore, high-sensitivity probes may not be required for security purposes. However, the following detection limits can be expected:

- High temperature method (between 680°C and 950°C or higher in a few special cases, best possible oxidation): 1 mg/L carbon

- Low temperature method (below 100°C, limited oxidation potential) = 0.2 mg/L carbon

The response time of a TOC analyzer may vary depending on the manufacturer's specifications, but it usually takes from 5 to 15 minutes to get a stable, accurate reading.

Chemical Sensors: Chlorine Measurement System

Residual chlorine is one of the most sensitive and useful indicator parameters in water distribution system monitoring. All water distribution systems monitor for residual chlorine concentrations as part of their Safe Drinking Water Act requirements, and procedures for monitoring chlorine concentrations are well established and accurate. Chlorine monitoring assures proper residual at all points in the system, helps pace rechlorination when needed, and quickly and reliably signals any unexpected increase in disinfectant demand. A significant decline or loss of residual chlorine could be an indication of potential threats to the system. Several key points regarding residual chlorine monitoring for security purposes are provided below:

- Chlorine residuals can be measured using continuous on-line monitors at fixed points in the system, or by taking grab samples at any point in the system and using chlorine test kits or portable sensors to determine chlorine concentrations.
- Correct placement of residual chlorine monitoring points within a system is crucial for early detection of potential threats. For example, while dead ends and low-pressure zones are common trouble spots that can show low residual chlorine concentrations, these zones are generally not of great concern for water security purposes because system hydraulics will limit the circulation of any contaminants present in these areas of the system.
- Monitoring point and monitoring procedures for SDWA compliance may be different from those for system security purposes, and utilities must determine the best use of on-line, fixed monitoring systems and portable sensors/test kits to balance their SDWA compliance and security needs.

Various portable and on-line chlorine monitors are commercially available. These range from sophisticated on-line chlorine monitoring systems to portable electrode sensors to colorimetric test kits. On-line systems can be equipped with control, signal, and alarm systems that notify the operator of low chlorine concentrations, and some may be tied into feedback loops that automatically adjust chlorine concentrations in the system. In contrast, the use of portable sensors or colorimetric test kits requires technicians to take a sample and read the results. The technician then initiates the required actions based on the results of the test.

Several measurement methods are currently available for measuring chlorine in water samples, including:

- N, N-diethyl-p-phenylenediamine (DPD) colorimetric method
- Iodometric method

- Amperometric electrodes
- Polarographic membrane sensors

It should be noted that there can be differences in the specific type of analyte, the range, and the accuracy of these different measurement methods. In addition, these different methods have different operations and maintenance requirements. For example, DPD systems require periodic replenishment of buffers, whereas polarographic systems do not. Users may want to consider these requirements when choosing the appropriate sensor for their system.

Chemical Sensors: Portable Cyanide Analyzer

Portable cyanide detection systems are designed to be used in the field to evaluate for potential cyanide contamination of a water asset. These detection systems use one of two distinct analytical methods—either a colorimetric method or an ion selective method—to provide a quick, accurate cyanide measurement that does not require laboratory evaluation. Aqueous cyanide chemistry can be complex. Various factors, including the water asset's pH and redox potential, can affect the toxicity of cyanide in that asset. While personnel using these cyanide detection devices do not need to have advanced knowledge of cyanide chemistry to successfully screen a water asset for cyanide, understanding aqueous cyanide chemistry can help users to interpret whether the asset's cyanide concentration represents a potential threat. Therefore, a short summary of aqueous cyanide chemistry, including a discussion of cyanide toxicity, is provided below. For more information, the reader is referred to Greenberg et al. (1999).

Cyanide (CN-) is a toxic carbon-nitrogen organic compound that is the functional portion of the lethal gas hydrogen cyanide (HCN). The toxicity of aqueous cyanide varies depending on its form. At near-neutral pH, "free cyanide" (which is commonly designated as "CN-," although it is actually defined as the total of HCN and CN-) is the predominant cyanide form in water. Free cyanide is potentially toxic in its aqueous form, although the primary concern regarding aqueous cyanide is that it could volatilize. Free cyanide is not highly volatile (it is less volatile than most volatile organic compounds (VOCs), but its volatility increases as the pH decreases below 8). However, when free cyanide does volatilize, it volatilizes in its highly toxic gaseous form (gaseous HCN). As a general rule, metal-cyanide complexes are much less toxic than free cyanide because they do not volatilize unless the pH is low.

Analyses for cyanide in public water systems are often conducted in certified labs using various USEPA-approved methods, such as the preliminary distillation procedure with subsequent analysis by a colorimetric, ion selective electrode, or flow injection methods. Lab analyses using these methods require careful sample preservation and pretreatment procedures and are generally expensive and time consuming. Using these methods, several cyanide fractions are typically defined:

- Total cyanide—includes free cyanide (CN- + HCN) and all metal-completed cyanide
- Weak Acid Dissociable (WAD) Cyanide—includes free cyanide (CN- + HCN) and weak cyanide complexes that could be potentially toxic by hydrolysis to free cyanide in the pH range 4.5–6.0

- Amendable Cyanide—includes free cyanide (CN- + HCN) and weak cyanide complexes that can release free cyanide at high pH (11–12) (this fraction gets its name because it includes measurement of cyanide from complexes that are "amendable" to oxidation by chlorine at high pH). To measure "Amenable Cyanide," the sample is split into two fractions. One of the fractions is analyzed for "Total Cyanide" as above. The other fraction is treated with high levels of chlorine for approximately one hour, dechlorinated, and distilled per the above "Total Cyanide" method. "Amendable Cyanide" is determined by the difference in the cyanide concentrations in these to fractions
- Soluble Cyanide—measures only soluble cyanide. Soluble cyanide is measured by using the preliminary filtration step, followed by "Total Cyanide" analysis described above

As discussed above, these different methods yield various different cyanide measurements which may or may not give a complete picture of that sample's potential toxicity. For example, the "Total Cyanide" method includes cyanide complexed with metals, some of which will not contribute to cyanide toxicity unless the pH is out of the normal range. In contrast, the "WAD Cyanide" measurement includes metal-complexed cyanide that could become free cyanide at low pH, and "Amendable Cyanide" measurements include metal-complexed cyanide that could become free cyanide at high pH. Personnel using these kits should therefore be aware of the potential differences in actual cyanide toxicity versus the cyanide potential differences in actual cyanide toxicity versus the cyanide measured in the sample under different environmental conditions.

Ingestion of aqueous cyanide can result in numerous adverse health effects and may be lethal. The USEPA's maximum contamination level for cyanide in drinking water is 0.2 µg/L (0.2 parts per million, or ppm). This MCL is based on free cyanide analysis per the "Amendable Cyanide" method described above (the USEPA has recognized that very stable metal-cyanide complexes such as iron-cyanide complex are nontoxic [unless exposed to significant UV radiation], and these fractions are therefore not considered when defining cyanide toxicity). Ingestion of free cyanide at concentrations in excess of this MCL causes both acute effects (e.g., rapid breathing, tremors and neurological symptoms) and chronic effects (e.g., weight loss, thyroid effects, and nerve damage). Under the current primary drinking water standards, public water systems are required to monitor their systems to minimize public exposure to cyanide levels in excess of the MCL.

Hydrogen cyanide gas is also toxic, and the Office of Safety and Health administration (OSHA) has set a permissible exposure limit (PEL) of 10 ppmv (parts per million by volume) for HCN inhalation. HCN also has a strong, bitter, almond-like smell and an odor threshold of approximately 1 ppmv. Considering the fact that HCN is relatively non-volatile (see above), a slight cyanide odor emanating from a water sample suggests very high aqueous cyanide concentrations—greater than 10–50 mg/L, which is in the range of a lethal or near lethal dose with the ingestion of one pint of water.

Portable Field Monitors to Measure VOCs

Volatile organic compounds are a group of highly utilized chemicals that have widespread applications, including use as fuel components, as solvents, and as cleaning and liquefying agents in degreasers, polishes, and dry cleaning solutions. VOCs are also used in herbicides and insecticides for agriculture applications. Laboratory-based methods for analyzing VOCs are well established; however, analyzing VOCs in the lab is time consuming—obtaining a result may require from several hours to several weeks depending on the specific method. Faster, commercially available methods for analyzing VOCs quickly in the field include use of portable gas chromatographs (GC), mass spectrometer (MS), or gas chromatographs/mass spectrometers (GC/MS), all of which can be used to obtain VOC concentration results within minutes. These instruments can be useful in rapid confirmation of the presence of VOCs in an asset, or for monitoring an asset on a regular basis. In addition, portable VOC analyzers can analyze for a wide range of VOCs, such as toxic industrial chemicals (TICs), chemical warfare agents (CWAs), drugs, explosives, and aromatic compounds. There are several easy-to-use, portable VOC analyzers currently on the market that are effective in evaluating VOC concentrations in the field. These instruments utilize gas chromatography, mass spectroscopy, or a combination of both methods, to provide near laboratory-quality analysis for VOCs.

Radiation Detection Equipment

Radioactive substances (radionuclides) are known health hazards that emit energetic waves and/or particles that can cause both carcinogenic and non-carcinogenic health effects. Radionuclides pose unique threats to source water supplies and water treatment, storage, or distribution systems because radiation emitted from radionuclides in water systems can affect individuals through several pathways—by direct contact with, ingestion or inhalation of, or external exposure to the contaminated water. While radiation can occur naturally in some cases due to the decay of some minerals, intentional and non-intentional releases of man-made radionuclides into water systems is also a realistic threat.

Threats to water and wastewater facilities from radioactive contamination could involve two major scenarios. First, the facility or its assets could be contaminated, preventing workers from accessing and operating the facility/assets. Second, at drinking water facilities, the water supply could be contaminated, and tainted water could be distributed to users downstream. These two scenarios require different threat reduction strategies. The first scenario requires that facilities monitor for radioactive substances being brought on-site; the second requires that water assets be monitored for radioactive contamination. While the effects of radioactive contamination are basically the same under both threat types, each of these threats requires different types of radiation monitoring and different types of equipment.

Radiation Detection Equipment for Monitoring Water Assets

Most water systems are required to monitor for radioactivity and certain radionuclides, and to meet maximum contaminant levels for these contaminants, to comply with the Safe Drinking Water Act. Currently, the USEPA requires drinking water to meet MCLs for beta/photon emitters (includes gamma radiation), alpha particles, combined radium 226/228, and uranium. However, this monitoring is required only at entry points into the system. In addition, after the initial sampling requirements, only one sample is required every three to nine years, depending on the contaminant type and the initial concentrations. While this is adequate to monitor for long-term protection from overall radioactivity and specific radionuclides in drinking water, it may not be adequate to identify short-term spikes in radioactivity, such as from spills, accidents, or intentional releases. In addition, compliance with the SDWA requires analyzing water samples in a laboratory, which results in a delay in receiving results. In contrast, security monitoring is more effective when results can be obtained quickly in the field. In addition, monitoring for security purposes does not necessarily require that the specific radionuclides causing the contamination be identified. Thus, for security purposes, it may be more appropriate to monitor for non-radionuclide-specific radiation using either portable field meters, which can be used as necessary to evaluate grab samples, or on-line systems, which can provide continuous monitoring of a system.

Ideally, measuring radioactivity in water assets in the field would involve minimal sampling and sample preparation. However, the physical properties of specific types of radiation combined with the physical properties of water make evaluating radioactivity in water assets in the field somewhat difficult. For example, alpha particles can only travel short distances and they cannot penetrate through most physical objects. Therefore, instruments designed to evaluate alpha emissions must be specially designed to capture emissions at a short distance from the source, and they must not block alpha emissions from entering the detector. Gamma radiation does not have the same types of physical properties, and thus it can be measured using different detectors.

Measuring different types of radiation is further complicated by the relationship between the radiation's intrinsic properties and the medium in which the radiation is being measured. For example, gas-flow proportional counters are typically used to evaluate gross alpha and beta radiation from smooth, solid surfaces, but due to the fact that water is not a smooth surface, and because alpha and beta emissions are relatively short range and can be attenuated within the water, these types of counters are not appropriate for measuring alpha and beta activity in water. An appropriate method for measuring alpha and beta radiation in water is by using a liquid scintillation counter. However, this requires mixing an aliquot of water with a liquid scintillation "cocktail." The liquid scintillation counter is a large, sensitive piece of equipment, so it is not appropriate for field use. Therefore, measurements for alpha and beta radiation from water assets are not typically made in the field.

Unlike the problems associated with measuring alpha and beta activity in water in the field, the properties of gamma radiation allow it to be measured relatively well in water samples in the field. The standard instrumentation used to measure gamma radiation from water samples in the field is a sodium iodide (NaI) scintillator.

Although the devices outlined above are the most commonly used for evaluating total alpha, beta, and gamma radiation, other methods and other devices can be used. In addition, local conditions (i.e., temperature, humidity, etc.) or the properties of the specific radionuclides emitting the radiation may make other types of devices or other methods more optimal to achieve the goals of the survey than the devices noted above. Therefore, experts or individual vendors should be consulted to determine the appropriate measurement device for any specific application.

An additional factor to consider when developing a program to monitor for radioactive contamination in water assets is whether to take regular grab samples or sample continuously. For example, portable sensors can be used to analyze grab samples at any point in the system, but have the disadvantage that they provide measurements only at one point in time. On the other hand, fixed-location sensors are usually used as part of a continuous, on-line monitoring system. These systems continuously monitor a water asset, and could be outfitted with some type of alarm system that would alert operators if radiation increased above a certain threshold. However, the sampling points are fixed and only certain points in the system can be monitored. In addition, the number of monitoring locations needed to capture the physical and radioactive complexity of a system can be prohibitive.

Toxicity Monitoring/Toxicity Meters

Toxicity measurement devices measure general toxicity to biological organisms, and detection of toxicity in any water/wastewater asset can indicate a potential threat, to the treatment process (in the case of influent toxicity), to human health (in the case of finished the of drinking water toxicity), or to the environment (in the case of effluent toxicity). Currently, whole effluent toxicity tests (WET tests), in which effluent samples are tested against test organisms, are required of many National Pollutant Discharge Elimination System discharge permits. The WET tests are used as a complement to the effluent limits on physical and chemical parameters to assess the overall effects of the discharge on living organisms or aquatic biota. Toxicity tests may also be used to monitor wastewater influent streams for potential hazardous contamination, such as organic heavy metals (arsenic, mercury, lead, chromium, and copper) that might upset the treatment process.

The ability to get feedback on sample toxicity from short-term toxicity tests or toxicity "meters" can be valuable in estimating the overall toxicity of a sample. On-line real-time toxicity monitoring is still under active research and development. However, there are several portable toxicity measurement devices commercially available. They can generally be divided into categories based on the different ways they measure toxicity:

- Meters measuring direct biological activity (e.g., luminescent bacteria) and correlating decreases in this direct biological activity with increased toxicity
- Meters measuring oxygen consumption and correlating decrease in oxygen consumption with increased toxicity.

COMMUNICATION AND INTEGRATION

This section discusses those devices necessary for communication and integration of water and wastewater system operations, such as electronic controllers, two-way radios, and wireless data communications. Electronic controllers are used to

Plant Security

automatically activate equipment (such as lights, surveillance cameras, audible alarms, or locks) when they are triggered. Triggering could be in response to a variety of scenarios, including tripping of an alarm or a motion sensor; breaking of a window or a glass door; variation in vibration sensor readings; or simply through input from a timer. Two-way wireless radios allow two or more users that have their radios tuned to the same frequency to communicate instantaneously with each other without the radios being physically lined together with wires or cables. Wireless data communications devices are used to enable transmission of data between computer systems and/or between a SCADA server and its sensing devices, without individual components being physically linked together via wires or cables. In water and wastewater utilities, these devices are often used to link remote monitoring stations (i.e., SCADA components) or portable computers (i.e., laptops) to computer networks without using physical wiring connections.

Electronic Controllers

An electronic controller is a piece of electronic equipment that receives incoming electric signals and uses preprogrammed logic to generate electronic output signals based on the incoming signals. While electronic controllers can be implemented for any application that involves inputs and outputs (for example, control of a piece of machinery in a factor), in a security application, these controllers essentially act as the system's "brain," and can respond to specific security-related inputs with preprogrammed output response. These systems combine the control of electronic circuitry with a logic function such that circuits are opened and closed (and thus equipment is turned on and off) through some preprogrammed logic. The basic principle behind the operation of an electrical controller is that it receives electronic inputs from sensors or any device generating an electrical signal (for example, electrical signals from motion sensors), and then uses preprogrammed logic to produce electrical outputs (for example, these outputs could turn on power to a surveillance camera or to an audible alarm). Thus, these systems automatically generate a preprogrammed, logical response to a preprogrammed input scenario.

The three major types of electronic controllers are timers, electromechanical relays, and programmable logic controllers (PLCs), which are often called "digital relays." Each of these types of controllers is discussed in more detail below. Timers use internal signal/inputs (in contrast to externally generated inputs) and generate electronic output signals at certain times. More specifically, timers control electric current flow to any application to which they are connected, and can turn the current on or off on a schedule pre-specified by the user. Typical timer range (amount of time that can be programmed to elapse before the timer activates linked equipment) is from 0.2 seconds to 10 hours, although some of the more advanced timers have ranges of up to 60 hours. Timers are useful in fixed applications that don't require frequent schedule changes. For example, a timer can be used to turn on the lights in a room or building at a certain time every day. Timers are usually connected to their own power supply (usually 120–240 V).

In contrast to timers, which have internal triggers based on a regular schedule, electromechanical relays and PLCs have both external inputs and external outputs. However, PLCs are more flexible and more powerful than are electromechanical relays, and thus this section focuses primarily on PLCs as the predominant technology for security-related electronic control applications. Electromechanical relays are simple devices that use a magnetic field to control a switch. Voltage applied to the relay's input coil creates a magnetic field, which attracts an internal metal switch. This causes the relay's contacts to touch, closing the switch and completing the electrical circuit. This activates any linked equipment. These types of systems are often used for high voltage applications, such as in some automotive and other manufacturing processes.

Two-Way Radios
Two-way radios, as discussed here, are limited to a direct unit-to-unit radio communication, either via single unit-to-unit transmission and reception or via multiple handheld units to a base station radio contact and distribution system. Radio frequency spectrum limitations apply to all handheld units, and directed by the Federation Communication Commission (FCC). This also distinguishes a handheld unit from a base station or base station unit (such as those used by an amateur (ham) radio operator), which operate under different wave length parameters.

Two-way radios allow a user to contact another user or group of users instantly on the same frequency, and to transmit voice or data without the need for wires. They use "half-duplex" communications, or communication that can be only transmitted or received but cannot transmit and receive simultaneously. In other words, only one person may talk, while other personnel with radio(s) can only listen. To talk, the user depresses the talk button and speaks into the radio. The audio then transmits the voice wirelessly to the receiving radios. When the speaker has finished speaking and the channel has cleared, users on any of the receiving radios can transmit; either to answer the first transmission or to begin a new conversation. In addition to carrying voice data, many types of wireless radios also allow the transmission of digital data, and these radios may be interfaced with computer networks that can use or track these data. For example, some two-way radios can send information such as global positioning system (GPS) data, or the ID of the radio. Some two-way radios can also send data through a SCADA system.

Wireless radios broadcast these voice or data communications over the airwaves from the transmitter to the receiver. While this can be an advantage in that the signal emanates in all directions and does not need a direct physical connection to be received at the receiver, it can also make the communications vulnerable to being blocked, intercepted, or otherwise altered. However, security features are available to ensure that the communications are not tampered with.

Wireless Data Communications
A wireless data communication system consists of two components: a "Wireless Access Point" (WAP) and a "Wireless Network Interface Card" (sometimes also referred to as a "Client"), which work together to complete the communications link. These wireless systems can link electronic devices, computers, and computer systems together using radio waves, thus eliminating the need for these individual components to be directly connected together through physical wires. While wireless data communications have widespread application in water and wastewater systems,

they also have limitations. First, wireless data connections are limited by the distance between components (radio waves scatter over a long distance and cannot be received efficiently, unless special directional antenna are used). Second, these devices only function if the individual components are in direct line of sight with each other, since radio waves are affected by interference from physical obstructions. However, in some cases, repeater units can be used to amplify and retransmit wireless signals to circumvent these problems. The two components of wireless devices are discussed in more detail below.

The wireless access point provides the wireless data communication service. It usually consists of a housing (which is constructed from plastic or metal depending on the environment it will be used in) containing a circuit board; flash memory that holds software; one of two external ports to connect to existing wired networks; a wireless radio transmitter/receiver; and one or more antenna connections. Typically, the WAP requires a one-time user configuration to allow the device to interact with the local area network (LAN). This configuration is usually done via a web-driven software application which is accessed via a computer.

Wireless network interface card or client is a piece of hardware that is plugged in to a computer and enables that computer to make a wireless network connection. The card consists of a transmitter, functional circuitry, and a receiver for the wireless signal, all of which work together to enable communication between the computer, its wireless transmitter/receiver, and its antenna connection. Wireless cards are installed in a computer through a variety of connections, including USB adapters, or Laptop CardBus (PCMCIA) or Desktop Peripheral (PCI) cards. As with the WAP, software is loaded onto the user's computer, allowing configuration of the card so that it may operate over the wireless network

Two of the primary applications for wireless data communications systems are to enable mobile or remote connections to a LAN, and to establish wireless communications links between SCADA remote telemetry units (RTUs) and sensors in the field. Wireless car connections are usually used for LAN access from mobile computers. Wireless cards can also be incorporated into RTUs to allow them to communicate with sensing devices that are located remotely.

CYBER PROTECTION DEVICES

Various cyber protection devices are currently available for use in protecting utility computer systems. These protection devices include antivirus and pest eradication software, firewalls, and network intrusion hardware/software. These products are discussed in this section.

Antivirus and Pest Eradication Software

Antivirus programs are designed to detect, delay, and respond to programs or pieces of code that are specifically designed to harm computers. These programs are known as "malware." Malware can include computer viruses, worms, and Trojan horse programs (programs that appear to be benign but which have hidden harmful effects). Pest eradication tools are designed to detect, delay, and respond to "spyware" (strategies that websites use to track user behavior, such as by sending "cookies" to the user's computer), and hacker tools that track keystrokes (keystroke loggers) or passwords (password crackers).

Viruses and pests can enter a computer system through the Internet or through infected floppy discs or CDs. They can also be placed onto a system by insiders. Some of these programs, such as viruses and worms, then move within a computer's drives and files, or between computers if the computers are networked to each other. This malware can deliberately damage files, utilize memory and network capacity, crash application programs, and initiate transmissions of sensitive information from a PC. While the specific mechanisms of these programs differ, they can infect files and even the basic operating program of the computer firmware/hardware.

The most important features of an antivirus program are its abilities to identify potential malware and to alert a user before infection occurs, as well as its ability to respond to a virus already resident on a system. Most of these programs provide a log so that the user can see what viruses have been detected and where they were detected. After detecting a virus, the antivirus software may delete the virus automatically, or it may prompt the user to delete the virus. Some programs will also fix files or programs damaged by the virus.

Various sources of information are available to inform the general public and computer system operators about new viruses being detected. Since antivirus programs use signatures (or snippets of code or data) to detect the presence of a virus, periodic updates are required to identify new threats. Many antivirus software providers offer free upgrades that are able to detect and respond to the latest viruses.

Firewalls

A firewall is an electronic barrier designed to keep computer hackers, intruders, or insiders from accessing specific data files and information on a utility's computer network or other electronic/computer systems. Firewalls operate by evaluating and then filtering information coming through a public network (such as the Internet) into the utility's computer or other electronic system. This evaluation can include identifying the source or destination addresses and ports, and allowing or denying access based on this identification. Two methods are used by firewalls to limit access to the utility's computers or other electronic systems from the public network:

- The firewall may deny all traffic unless it meets certain criteria.
- The firewall may allow all traffic through unless it meets certain criteria.

A simple example of the first method is to screen requests to ensure that they come from an acceptable (i.e., previously identified) domain name and Internet protocol address. Firewalls may also use more complex rules that analyze the application data to determine if the traffic should be allowed through. For example, the firewall may require user authentication (i.e., use of a password) to access the system. How a firewall determines what traffic to let through depends on which network layer it operates at and how it is configured. Firewalls may be a piece of hardware, a software program, or an appliance card that contains both.

Advanced features that can be incorporated into firewalls allow for the tracking of attempts to log-on to the local area network system. For example, a report of successful and unsuccessful log-in attempts may be generated for the computer specialist to analyze. For systems with mobile users, firewalls allow remote access to the private network by the use of secure log-on procedures and authentication certificates. Most firewalls have a graphical user interface for managing the firewall. In addition, new Ethernet firewall cards that fit in the slot of an individual computer bundle provide additional layers of defense (like encryption and permit/deny) for individual computer transmissions to the network interface function. The cost of these new cards is only slightly higher than traditional network interface cards.

Network Intrusion Hardware and Software

Network intrusion detection and prevention system are software- and hardware-based programs designed to detect unauthorized attacks on a computer network system. Whereas other applications such as firewalls and antivirus software share similar objectives with network intrusion systems, network intrusion systems provide a deeper layer of protection beyond the capabilities of these other systems because they evaluate pattern of computer activity rather than specific files. It is worth noting that attacks may come from either outside or within the system (i.e., from an insider), and that network intrusion detection systems may be more applicable for detecting patterns of suspicious activity from inside a facility (i.e., accessing sensitive data, etc.) than are other information technology solutions. Network intrusion detection systems employ a variety of mechanisms to evaluate potential threats. The types of search and detection mechanisms are dependent upon the level of sophistication of the system. Some of the available detection methods include:

- Protocol analysis—this is the process of capturing, decoding, and interpreting electronic traffic. The protocol analysis method of network intrusion detection involves the analysis of data captured during transactions between two or more systems or devices, and the evaluation of these data to identify unusual activity and potential problems. Once a problem is isolated and recorded, problems or potential threats can be linked to pieces of hardware or software. Sophisticated protocol analysis will also provide statistics and trend information on the captured traffic.
- Traffic anomaly detection—it identifies potential threatening activity by comparing incoming traffic to "normal" traffic patterns, and identifying deviations. It does this by comparing user characteristics against thresholds and triggers defined by the network administrator. This method is designed to detect attacks that span a number of connections, rather than a single session.
- Network honeypot—this method establishes nonexistent services in order to identify potential hackers. A network honeypot impersonates services that don't exist by sending fake information to people scanning the network. It identifies the attacker when they attempt to connect to the service. There is no reason for legitimate traffic to access these resources because they don't exist; therefore any attempt to access them constitutes an attack.

- Anti-intrusion detection system evasion techniques—these methods are designed to protect against attackers who may be trying to evade intrusion detection system scanning. They include methods called IP defragmentation, TCP streams reassembly, and deobfuscation.

These detection systems are automated, but they can only indicate patterns of activity, and a computer administrator or other experienced individual must interpret activities to determine whether or not they are potentially harmful. Monitoring the logs generated by these systems can be time consuming, and there may be a learning curve to determine a baseline of "normal" traffic patterns from which to distinguish potential suspicious activity.

Adoption of Technologies with Known Vulnerabilities

When a technology is not well known, not widely used, not understood or publicized, it is difficult to penetrate it and thus disable it. Historically, proprietary hardware, software, and network protocols made it difficult to understand how control systems operated—and therefore how to hack into them. Today, however, to reduce costs and improve performance, organizations have been transitioning from proprietary systems to less expensive, standardized technologies such as Microsoft's Windows and Unix-like operating systems and the common networking protocols used by the Internet. These widely used standardized technologies have commonly known vulnerabilities, and sophisticated and effective exploitation tools are widely available and relatively easy to use. As a consequence, both the number of people with the knowledge to wage attacks and the number of systems subject to attack have increased. Also, common communication protocols and the emerging use of Extensible Markup Language (commonly referred to as XML) can make it easier for a hacker to interpret the content of communications among the components of a control system.

Control systems are often connected to other networks—enterprises often integrate their control system with their enterprise networks. This increased connectivity has significant advantages, including providing decision makers with access to real-time information and allowing engineers to monitor and control the process control system from different points on the enterprise network. In addition, enterprise networks are often connected to the networks of strategic partners and to the Internet. Further, control systems are increasingly using wide area networks and the Internet to transmit data to their remote or local stations and individual devices. This convergence of control networks with public and enterprise networks potentially exposes the control systems to additional security vulnerabilities. Unless appropriate security controls are deployed in the enterprise network and the control system network, breaches in enterprise security can affect the operation of control systems. According to industry experts, the use of existing security technologies, as well as strong user authentication and patch management practices, are generally not implemented in control systems because control systems operate in real time, typically are not designed with cyber security in mind, and usually have limited processing capabilities.

Existing security technologies such as authorization, authentication, encryption, intrusion detection, and filtering of network traffic and communications require more bandwidth, processing power, and memory than control system components typically have. Because controller stations are generally designed to do specific tasks, they use low-cost, resource-constrained microprocessors. In fact, some devices in the electrical industry still use the Intel 8088 processor, introduced in 1978. Consequently, it is difficult to install existing security technologies without seriously degrading the performance of the control system.

Further, complex passwords and other strong password practices are not always used to prevent unauthorized access to control systems, in part because this could hinder a rapid response to safety procedures during an emergency. As a result, according to experts weak passwords that are easy to guess, shared, and infrequently changed, including the use of default passwords or even no password at all, are reportedly common in control systems,.

In addition, although modern control systems are based on standard operating systems, they are typically customized to support control system applications. Consequently, vendor-provided software patches are generally incompatible or cannot be implemented without compromising service shutting down "always-on" systems or affecting interdependent operations.

Potential vulnerabilities in control systems are exacerbated by insecure connections. Organizations often leave access links—such as dial-up modems to equipment and control information—open for remote diagnostics, maintenance, and examination of system status. Such links may not be protected with authentication of encryption, which increases the risk that hackers could us these insecure connections to break into remotely controlled systems. Also, control systems often use wireless communications systems, which are especially vulnerable to attack, or leased lines that pass through commercial telecommunications facilities. Without encryption to protect data as it flows through these insecure connections or authentication mechanisms to limit access, there is limited protection for the integrity of the information being transmitted.

Public information about infrastructures and control systems is available to potential hackers and intruders. The availability of this infrastructure and vulnerability data was demonstrated by a university graduate student, whose dissertation reportedly mapped every business and industrial sector in the American economy to the fiber optic network that connects them—using material that was available publicly on the Internet, none of which was classified. Many of the electric utility officials who were interviewed for the National Security Telecommunications Advisory Committee's Information Assurance Task Force's Electric Power Risk Assessment expressed concern over the amount of information about their infrastructure that is readily available to the public.

In the electric power industry, open sources of information—such as product data and educational videotapes from engineering associations—can be used to understand the basics of the electrical grid. Other publicly available information—including filings of the Federal Energy Regulatory Commission (FERC), industry publications, maps, and material available on the Internet—is sufficient to allow someone to identify the most heavily loaded transmission lines and the most critical substations in the power grid.

In addition, significant information on control systems is publicly available—including design and maintenance documents, technical standards for the interconnection of control systems and RTUs, and standards for communication among control devise—all of which could assist hackers in understanding the systems and how to attack them. Moreover, there are numerous former employees, vendor, support contractors, and other end users of the same equipment worldwide with inside knowledge of the operation of control systems.

CYBER THREATS TO CONTROL SYSTEMS

There is a general consensus—and increasing concern—among government officials and experts on control systems about potential cyber threats to the control systems that govern our critical infrastructures. As components of control systems increasingly make critical decisions that were once made by humans, the potential effect of a cyber threat becomes more devastating. Such cyber threats could come from numerous sources, ranging from hostile governments and terrorist groups to disgruntled employees and other malicious intruders. Based on interviews and discussions with representatives throughout the electric power industry, the Information Assurance Task Force of the National Security Telecommunications Advisory Committee concluded that an organization with sufficient resources, such as a foreign intelligence service or a well-supported terrorist group, could conduct a structured attack on the electric power grid electronically, with a high degree of anonymity and without having to set foot in the target nation.

In July 2002, National Infrastructure Protection Center (NIPC) reported that the potential for compound cyber and physical attacks, referred to as "swarming attacks," is an emerging threat to the U.S. critical infrastructure. As NIPC reports, the effects of a swarming attack include slowing or complicating the response to a physical attack. For instance, a cyber attack that disabled the water supply or the electrical system in conjunction with a physical attack could deny emergency services the necessary resources to manage the consequences—such as controlling fires, coordinating actions, and generating light.

Control systems, such as SCADA, can be vulnerable to cyber attacks. Entities or individuals with malicious intent might take one or more of the following actions to successfully attack control systems:

- Disrupt the operation of control systems by delaying or blocking the flow of information through control networks, thereby denying availability of the networks to control system operations
- Make unauthorized changes to programmed instructions in PLCs, RTUs, or DCS controllers, change alarm thresholds, or issue unauthorized commands to control equipment, which could potentially result in damage to equipment (if tolerances are exceeded), premature shutdown of processes (such as prematurely shutting down transmission lines), or even disabling of control equipment
- Send false information to control system operators either to disguise unauthorized changes or to initiate inappropriate actions by system operators

- Modify the control system software, producing unpredictable results
- Interfere with the operation of safety systems

In addition, in control systems that cover a wide geographic area, the remote sites are often unstaffed and may not be physically monitored. If such remote systems are physically breached, the attackers could establish a cyber connection to the control network.

SECURING CONTROL SYSTEMS

Several challenges must be addressed to effectively secure control systems against cyber threats. These challenges include: (1) the limitations of current security technologies in securing control systems; (2) the perception that securing control systems may not economically justifiable; and (3) the conflicting priorities within organizations regarding the security of control systems. A significant challenge in effectively securing control systems is the lack of specialized security technologies for these systems. The computing resources in control systems that are needed to perform security functions tend to be quite limited, making it very difficult to use security technologies within control system networks without severely hindering performance. Securing control systems may not be perceived as economically justifiable. Experts and industry representatives have indicated that organizations may be reluctant to spend more money to secure control systems. Hardening the security of control systems would require industries to expend more resources, including acquiring more personnel, providing training for personnel, and potentially prematurely replacing current systems that typically have a lifespan of about 20 years. Finally, several experts and industry representatives indicated that the responsibility for securing control systems typically includes two separate groups: IT security personnel and control system engineers and operators. IT security personnel tend to focus on securing enterprise systems, while control system engineers and operators tend to be more concerned with the reliable performance of their control systems. Further, they indicate that, as a result, those two groups do not always fully understand each other's requirements and do not collaborate to implement secure control systems.

REFERENCES AND RECOMMENDED READING

DOE. 2001. *21 Steps to Improve Cyber Security of SCADA Networks.* Washington, DC: Department of Energy.
Ezell, B.C. 1998. *Risks of Cyber Attack to Supervisory Control and Data Acquisition.* Charlottesville, VA: University of Virginia.
FBI. 2000. *Threat to Critical Infrastructure.* Washington, DC: Federal Bureau of Investigation.
FBI. 2004. *Ninth Annual Computer Crime and Security Survey.* FBI: Computer Crime Institute and Federal Bureau of Investigations, Washington, DC.
GAO. 2003. *Critical Infrastructure Protection: Challenges in Securing Control System.* Washington, DC: United States General Accounting Office.
Gellman, B. 2002. Cyber-Attacks by Al Qaeda Feared: Terrorists at Threshold of Using Internet as Tool of Bloodshed, Experts Say. *Washington Post*, June 27, p. A01.
Henry, K. 2002. New Face of Security. *Government Security*, April, pp. 30–31.

IBWA. 2004. *Bottled Water Safety and Security.* Alexandria, VA: International Bottled Water Association.
NIPC. 2002. *National Infrastructure Protection Center Report.* Washington, DC: National Infrastructure Protection Center.
Stamp, J., et al. 2003. *Common Vulnerabilities in Critical Infrastructure Control Systems.* 2nd ed. Sandia National Laboratories.
USEPA. 2004. *Water Security: Basic Information.* Accessed 30 September 2019 @ http://cfpub.epa.gov/ safewater/watersecurity/basicinformation.cfm.
USEPA. 2005. EPA Needs to Determine What Barriers Prevent Water Systems from Securing Known SCADA Vulnerabilities. In Harris, J. (ed.), *Final Briefing Report.* Washington, DC: USEPA.

Glossary

WASTEWATER AND DRINKING WATER TERMINOLOGY AND DEFINITIONS

To study any aspect of wastewater and drinking water treatment operations, you must master the language associated with the technology. Each technology has its own terms with its own accompanying definitions. Many of the terms used in water/wastewater treatment are unique; others combine words from many different technologies and professions. One thing is certain: water/wastewater public utility managers without a clear understanding of the terms related to their profession are ill-equipped to perform their duties in the manner required. Thus, this glossary is actually an information source for public utility managers and serves as a quick reference—basically a dictionary—so that the manager can refer to terminology listed herein and be able to find and understand important definitions of important terms.

A

Absorb: to take in. Many things absorb water.

Acre-feet (acre-foot): an expression of water quantity. One acre-foot will cover one acre of ground one foot deep. An acre-foot contains 43,560 cubic feet, 1,233 cubic meters, or 325,829 gal (U.S). Also abbreviated as ac-ft.

Activated carbon: derived from vegetable or animal materials by roasting in a vacuum furnace. Its porous nature gives it a very high surface area per unit mass—as much as 1,000 square meters per gram, which is 10 million times the surface area of 1 gram of water in an open container. Used in adsorption (see definition), activated carbon adsorbs substances that are not or are only slightly adsorbed by other methods.

Activated sludge: the solids formed when microorganisms are used to treat wastewater using the activated sludge treatment process. It includes organisms, accumulated food materials, and waste products from the aerobic decomposition process.

Adsorption: the adhesion of a substance to the surface of a solid or liquid. Adsorption is often used to extract pollutants by causing them to attach to such adsorbents as activated carbon or silica gel. *Hydrophobic* water-repulsing adsorbents are used to extract oil from waterways in oil spills.

Advanced wastewater treatment: treatment technology to produce an extremely high quality discharge.

Aeration: the process of bubbling air through a solution, sometimes cleaning water of impurities by exposure to air.

Aerobic: conditions in which free, elemental oxygen is present. Also used to describe organisms, biological activity, or treatment processes that require free oxygen.

Agglomeration: floc particles colliding and gathering into a larger settleable mass.

Air gap: the air space between the free-flowing discharge end of a supply pipe and an unpressurized receiving vessel.

Algae bloom: a phenomenon whereby excessive nutrients within a river, stream, or lake causes an explosion of plant life that results in the depletion of the oxygen in the water needed by fish and other aquatic life. Algae bloom is usually the result of urban runoff (of lawn fertilizers, etc.). The potential tragedy is that of a "fish kill," where the stream life dies in one mass execution.

Alum: aluminum sulfate; a standard coagulant used in water treatment.

Ambient: the expected natural conditions that occur in water unaffected or uninfluenced by human activities.

Anaerobic: conditions in which no oxygen (free or combined) is available. Also used to describe organisms, biological activity, or treatment processes that function in the absence of oxygen.

Anoxic: conditions in which no free, elemental oxygen is present. The only source of oxygen is combined oxygen, such as that found in nitrate compounds. Also used to describe biological activity of treatment processes that function only in the presence of combined oxygen.

Aquifer: a water-bearing stratum of permeable rock, sand, or gravel.

Aquifer system: a heterogeneous body of introduced permeable and less permeable material that acts as a water-yielding hydraulic unit of regional extent.

Artesian water: a well tapping a confined or artesian aquifer in which the static water level stands above the top of the aquifer. The term is sometimes used to include all wells tapping confined water. Wells with water level above the water table are said to have positive artesian head (pressure) and those with water level below the water table, negative artesian head.

Average monthly discharge limitation: the highest allowable discharge over a calendar month.

Average weekly discharge limitation: the highest allowable discharge over a calendar week.

B

Backflow: reversal of flow when pressure in a service connection exceeds the pressure in the distribution main.

Backwash: fluidizing filter media with water, air, or a combination of the two so that individual grains can be cleaned of the material that has accumulated during the filter run.

Bacteria: any of a number of one-celled organisms, some of which cause disease.

Bar screen: a series of bars formed into a grid used to screen out large debris from influent flow.

Base: a substance that has a pH value between 7 and 14.

Basin: a groundwater reservoir defined by the overlying land surface and underlying aquifers that contain water stored in the reservoir.

Beneficial use of water: the use of water for any beneficial purpose. Such uses include domestic use, irrigation, recreation, fish and wildlife, fire protection, navigation, power, industrial use, etc. The benefit varies from one location to another and by custom. What constitutes beneficial use is often defined by statute or court decisions.

Biochemical oxygen demand (BOD_5): the oxygen used in meeting the metabolic needs of aerobic microorganisms in water rich in organic matter.

Biosolids: from *Merriam-Webster's Collegiate Dictionary, Tenth Ed.* (1998): biosolids *n* (1977) solid organic matter recovered from a sewage treatment process and used especially as fertilizer (or soil amendment)—usually used in plural.

Note: In this text, *biosolids* is used in many places to replace the standard term sludge. The author views the term sludge as an ugly four-letter word inappropriate to use to describe biosolids. Biosolids is a product that can be reused; it has some value. Because biosolids have value, they certainly should not be classified as a "waste" product—and when biosolids for beneficial reuse is addressed, it is made clear that it is not.

Biota: all the species of plants and animals indigenous to a certain area.

Boiling point: the temperature at which a liquid boils. The temperature at which the vapor pressure of a liquid equals the pressure on its surface. If the pressure of the liquid varies, the actual boiling point varies. The boiling point of water is 212° Fahrenheit or 100° Celsius.

Breakpoint: point at which chlorine dosage satisfies chlorine demand.

Breakthrough: in filtering, when unwanted materials start to pass through the filter.

Buffer: a substance or solution that resists changes in pH.

C

Calcium carbonate: compound principally responsible for hardness.

Calcium hardness: portion of total hardness caused by calcium compounds.

Carbonate hardness: caused primarily by compounds containing carbonate.

Carbonaceous biochemical oxygen demand, $CBOD_5$: the amount of biochemical oxygen demand that can be attributed to carbonaceous material.

Chemical oxygen demand (COD): the amount of chemically oxidizable materials present in the wastewater.

Chlorination: disinfection of water using chlorine as the oxidizing agent.

Clarifier: a device designed to permit solids to settle or rise and be separated from the flow. Also known as a settling tank or sedimentation basin.

Coagulation: the neutralization of the charges of colloidal matter.

Coliform: a type of bacteria used to indicate possible human or animal contamination of water.

Combined sewer: a collection system that carries both wastewater and stormwater flows.

Comminution: a process to shred solids into smaller, less harmful particles.

Composite sample: a combination of individual samples taken in proportion to flow.

Connate water: pressurized water trapped in the pore spaces of sedimentary rock at the time it was deposited. It is usually highly mineralized.

Consumptive use: (1) the quantity of water absorbed by crops and transpired or used directly in the building of plant tissue, together with the water evaporated from the cropped area. (2) The quantity of water transpired and evaporated from a cropped area or the normal loss of water from the soil by evaporation and plant transpiration. (3) The quantity of water discharged to the atmosphere or incorporated in the products of the process in connection with vegetative growth, food processing, or an industrial process.

Contamination (water): damage to the quality of water sources by sewage, industrial waste, or other material.

Cross-connection: a connection between a storm drain system and a sanitary collection system; a connection between two sections of a collection system to handle anticipated overloads of one system; or a connection between drinking (potable) water and an unsafe water supply or sanitary collection system.

D

Daily discharge: the discharge of a pollutant measured during a calendar day or any 24-hour period that reasonably represents a calendar day for the purposes of sampling. Limitations expressed as weight are total mass (weight) discharged over the day. Limitations expressed in other units are average measurement of the day.

Daily maximum discharge: the highest allowable values for a daily discharge.

Darcy's Law: an equation for the computation of the quantity of water flowing through porous media. Darcy's Law assumes that the flow is laminar and that inertia can be neglected. The law states that the rate of viscous flow of homogenous fluids through isotropic porous media is proportional to, and in the direction of, the hydraulic gradient.

Detention time: the theoretical time water remains in a tank at a given flow rate.

De-watering: the removal or separation of a portion of water present in a sludge or slurry.

Diffusion: the process by which both ionic and molecular species dissolved in water move from areas of higher concentration to areas of lower concentration.

Discharge monitoring report (DMR): the monthly report required by the treatment plant's National Pollutant Discharge Elimination System (NPDES) discharge permit.

Disinfection: water treatment process that kills pathogenic organisms.

Disinfection by-products (DBPs): chemical compounds formed by the reaction of disinfectant with organic compounds in water.

Dissolved oxygen (DO): the amount of oxygen dissolved in water or sewage. Concentrations of less than five parts per million (ppm) can limit aquatic life or cause offensive odors. Excessive organic matter present in water because of inadequate waste treatment and runoff from agricultural or urban land generally causes low DO.

Dissolved solids: the total amount of dissolved inorganic material contained in water or wastes. Excessive dissolved solids make water unsuitable for drinking or industrial uses.

Glossary

Domestic consumption (use): water used for household purposes such as washing, food preparation and showers. The quantity (or quantity per capita) of water consumed in a municipality or district for domestic uses or purposes during a given period, it sometimes encompasses all uses, including the quantity wasted, lost, or otherwise unaccounted for.

Drawdown: lowering the water level by pumping. It is measured in feet for a given quantity of water pumped during a specified period, or after the pumping level has become constant.

Drinking water standards: established by state agencies, U.S. Public Health Service, and Environmental Protection Agency (EPA) for drinking water in the United States.

E

Effluent: something that flows out, usually a polluting gas or liquid discharge.

Effluent limitation: any restriction imposed by the regulatory agency on quantities, discharge rates, or concentrations of pollutants discharged from point sources into state waters.

Energy: in scientific terms, the ability or capacity of doing work. Various forms of energy include kinetic, potential, thermal, nuclear, rotational, and electromagnetic. One form of energy may be changed to another, as when coal is burned to produce steam to drive a turbine, which produces electric energy.

Erosion: the wearing away of the land surface by wind, water, ice, or other geologic agents. Erosion occurs naturally from weather or runoff but is often intensified by human land use practices.

Eutrophication: the process of enrichment of water bodies by nutrients. Eutrophication of a lake normally contributes to its slow evolution into a bog or marsh and ultimately to dry land. Eutrophication may be accelerated by human activities, thereby speeding up the aging process.

Evaporation: the process by which water becomes a vapor at a temperature below the boiling point.

F

Facultative: organisms that can survive and function in the presence or absence of free, elemental oxygen.

Fecal coliform: the portion of the coliform bacteria group that is present in the intestinal tracts and feces of warm-blooded animals.

Field capacity: the capacity of soil to hold water. It is measured as the ratio of the weight of water retained by the soil to the weight of the dry soil.

Filtration: the mechanical process that removes particulate matter by separating water from solid material, usually by passing it through sand.

Floc: solids that join to form larger particles that will settle better.

Flocculation: slow mixing process in which particles are brought into contact, with the intent of promoting their agglomeration.

Flume: a flow rate measurement device.

Fluoridation: chemical addition to water to reduce incidence of dental caries in children.

Food-to-microorganisms ratio (F/M): an activated sludge process control calculation based upon the amount of food (BOD_5 or COD) available per pound of mixed liquor volatile suspended solids.

Force main: a pipe that carries wastewater under pressure from the discharge side of a pump to a point of gravity flow downstream.

G

Grab sample: an individual sample collected at a randomly selected time.

Graywater: water that has been used for showering, clothes washing, and faucet uses. Kitchen sink and toilet water are excluded. This water has excellent potential for reuse as irrigation for yards.

Grit: heavy inorganic solids, such as sand, gravel, eggshells, or metal filings.

Groundwater: the supply of fresh water found beneath the earth's surface (usually in aquifers) often used for supplying wells and springs. Because groundwater is a major source of drinking water, concern is growing over areas where leaching agricultural or industrial pollutants or substances from leaking underground storage tanks (USTs) are contaminating groundwater.

Groundwater hydrology: the branch of hydrology that deals with groundwater; its occurrence and movements, its replenishment and depletion, the properties of rocks that control groundwater movement and storage, and the methods of investigation and use of groundwater.

Groundwater recharge: the inflow to a groundwater reservoir.

Groundwater runoff: a portion of runoff that has passed into the ground, has become groundwater, and has been discharged into a stream channel as spring or seepage water.

H

Hardness: the concentration of calcium and magnesium salts in water.

Headloss: amount of energy used by water in moving from one point to another.

Heavy metals: metallic elements with high atomic weights, for example, mercury, chromium, cadmium, arsenic, and lead. They can damage living things at low concentrations and tend to accumulate in the food chain.

Holding pond: a small basin or pond designed to hold sediment laden or contaminated water until it can be treated to meet water quality standards or used in some other way.

Hydraulic cleaning: cleaning pipe with water under enough pressure to produce high water velocities.

Hydraulic gradient: a measure of the change in groundwater head over a given distance.

Hydraulic head: the height above a specific datum (generally sea level) that water will rise in a well.

Hydrologic cycle (water cycle): the cycle of water movement from the atmosphere to the earth and back to the atmosphere through various processes. These processes include: precipitation, infiltration, percolation, storage, evaporation, transpiration and condensation.

Hydrology: the science dealing with the properties, distribution, and circulation of water.

I

Impoundment: a body of water such as a pond, confined by a dam, dike, floodgate, or other barrier, and used to collect and store water for future use.

Industrial wastewater: wastes associated with industrial manufacturing processes.

Infiltration: the gradual downward flow of water from the surface into soil material.

Infiltration/inflow: extraneous flows in sewers; simply, inflow is water discharged into sewer pipes or service connections from such sources as foundation drains, roof leaders, cellar and yard area drains, cooling water from air conditioners, and other clean water discharges from commercial and industrial establishments. Defined by Metcalf & Eddy as follows:*Infiltration*—water entering the collection system through cracks, joints, or breaks.

- *Steady inflow*—water discharged from cellar and foundation drains, cooling water discharges, and drains from springs and swampy areas. This type of inflow is steady and is identified and measured along with infiltration.
- *Direct flow*—those types of inflow that have a direct stormwater runoff connection to the sanitary sewer and cause an almost immediate increase in wastewater flows. Possible sources are roof leaders, yard and areaway drains, manhole covers, cross-connections from storm drains and catch basins, and combined sewers.
- *Total inflow*—the sum of the direct inflow at any point in the system plus any flow discharged from the system upstream through overflows, pumping station bypasses, and the like.
- *Delayed inflow*—stormwater that may require several days or more to drain through the sewer system. This category can include the discharge of sump pumps from cellar drainage as well as the slowed entry of surface water through manholes in ponded areas.

Influent: wastewater entering a tank, channel, or treatment process.

Inorganic chemical/compounds: chemical substances of mineral origin, not of carbon structure. These include metals such as lead, iron (ferric chloride), and cadmium.

Ion exchange process: used to remove hardness from water.

J

Jar Test: laboratory procedure used to estimate proper coagulant dosage.

L

Langelier saturation index (L.I.): a numerical index that indicates whether calcium carbonate will be deposited or dissolved in a distribution system.

Leaching: the process by which soluble materials in the soil such as nutrients, pesticide chemicals, or contaminants are washed into a lower layer of soil or are dissolved and carried away by water.

License: a certificate issued by the State Board of Waterworks/Wastewater Works Operators authorizing the holder to perform the duties of a wastewater treatment plant operator.

Lift station: a wastewater pumping station designed to "lift" the wastewater to a higher elevation. A lift station normally employs pumps or other mechanical devices to pump the wastewater and discharges into a pressure pipe called a force main.

M

Maximum contaminant level (MCL): an enforceable standard for protection of human health.

Mean cell residence time (MCRT): the average length of time a mixed liquor suspended solids particle remains in the activated sludge process. May also be known as sludge retention time.

Mechanical cleaning: clearing pipe by using equipment (bucket machines, power rodders, or hand rods) that scrapes, cuts, pulls, or pushes the material out of the pipe.

Membrane process: a process that draws a measured volume of water through a filter membrane with small enough openings to take out contaminants.

Metering pump: a chemical solution feed pump that adds a measured amount of solution with each stroke or rotation of the pump.

Milligrams/liter (mg/L): a measure of concentration equivalent to parts per million (ppm).

Mixed liquor: the suspended solids concentration of the mixed liquor.

Mixed liquor volatile suspended solids (MLVSS): the concentration of organic matter in the mixed liquor suspended solids.

N

Nephelometric turbidity unit (NTU): indicates the amount of turbidity in a water sample.

Nitrogenous oxygen demand (NOD): a measure of the amount of oxygen required to biologically oxidize nitrogen compounds under specified conditions of time and temperature.

Nonpoint source (NPS) pollution: forms of pollution caused by sediment, nutrients, organic and toxic substances originating from land use activities that are carried to lakes and streams by surface runoff. Nonpoint source pollution occurs when the rate of materials entering these water bodies exceeds natural levels.

NPDES permit: National Pollutant Discharge Elimination System permit, which authorizes the discharge of treated wastes and specifies the conditions that must be met for discharge.

Nutrients: substances required to support living organisms. Usually refers to nitrogen, phosphorus, iron, and other trace metals.

O

Organic chemicals/compounds: animal or plant-produced substances containing mainly carbon, hydrogen, and oxygen, such as benzene and toluene.

P

Parts per million (ppm): the number of parts by weight of a substance per million parts of water. This unit is commonly used to represent pollutant concentrations. Large concentrations are expressed in percentages.

Pathogenic: disease causing. A pathogenic organism is capable of causing illness.

Percolation: the movement of water through the subsurface soil layers, usually continuing downward to the groundwater or water table reservoirs.

pH: a way of expressing both acidity and alkalinity on a scale of 0–14, with 7 representing neutrality; numbers less than 7 indicate increasing acidity and numbers greater than 7 indicate increasing alkalinity.

Photosynthesis: a process in green plants in which water, carbon dioxide and sunlight combine to form sugar.

Piezometric surface: an imaginary surface that coincides with the hydrostatic pressure level of water in an aquifer.

Point source pollution: a type of water pollution resulting from discharges into receiving waters from easily identifiable points. Common point sources of pollution are discharges from factories and municipal sewage treatment plants.

Pollution: the alteration of the physical, thermal, chemical, or biological quality of, or the contamination of, any water in the state that renders the water harmful, detrimental, or injurious to humans, animal life, vegetation, property, or to public health, safety, or welfare, or impairs the usefulness or the public enjoyment of the water for any lawful or reasonable purpose.

Porosity: that part of a rock that contains pore spaces without regard to size, shape, interconnection, or arrangement of openings. It is expressed as percentage of total volume occupied by spaces.

Potable water: water satisfactorily safe for drinking purposes from the standpoint of its chemical, physical, and biological characteristics.

Precipitate: a deposit on the earth of hail, rain, mist, sleet, or snow. The common process by which atmospheric water becomes surface or subsurface water, the term *Precipitation* is also commonly used to designate the quantity of water precipitated.

Preventive maintenance (PM): regularly scheduled servicing of machinery or other equipment using appropriate tools, tests, and lubricants. This type of maintenance can prolong the useful life of equipment and machinery and increase its efficiency by detecting and correcting problems before they cause a breakdown of the equipment.

Purveyor: an agency or person that supplies potable water.

R

Radon: a radioactive, colorless, odorless gas that occurs naturally in the earth. When trapped in buildings, concentrations build up, and can cause health hazards such as lung cancer.

Recharge: the addition of water into a groundwater system.

Reservoir: a pond, lake, tank, or basin (natural or human made) where water is collected and used for storage. Large bodies of groundwater are called groundwater reservoirs; water behind a dam is also called a reservoir of water.

Return activated sludge solids (RASS): the concentration of suspended solids in the sludge flow being returned from the settling tank to the head of the aeration tank.

Reverse osmosis: process in which almost pure water is passed through a semipermeable membrane.

River basin: a term used to designate the area drained by a river and its tributaries.

S

Sanitary wastewater: wastes discharged from residences and from commercial, institutional, and similar facilities that include both sewage and industrial wastes.

Schmutzdecke: layer of solids and biological growth that forms on top of a slow sand filter, allowing the filter to remove turbidity effectively without chemical coagulation.

Scum: the mixture of floatable solids and water removed from the surface of the settling tank.

Sediment: transported and deposited particles derived from rocks, soil, or biological material.

Sedimentation: a process that reduces the velocity of water in basins so that suspended material can settle out by gravity.

Seepage: the appearance and disappearance of water at the ground surface. Seepage designates movement of water in saturated material. It differs from percolation, which is predominantly the movement of water in unsaturated material.

Septic tanks: used to hold domestic wastes when a sewer line is not available to carry them to a treatment plant. The wastes are piped to underground tanks directly from a home or homes. Bacteria in the wastes decompose some of the organic matter, the sludge settles on the bottom of the tank, and the effluent flows out of the tank into the ground through drains.

Settleability: a process control test used to evaluate the settling characteristics of the activated sludge. Readings taken at 30–60 minutes are used to calculate the settled sludge volume (SSV) and the sludge volume index (SVI).

Settled sludge volume (SSV): the volume (in percent) occupied by an activated sludge sample after 30–60 minutes of settling. Normally written as SSV with a subscript to indicate the time of the reading used for calculation (SSV_{60} or SSV_{30}).

Sludge: the mixture of settleable solids and water removed from the bottom of the settling tank.

Sludge retention time (SRT): see mean cell residence time.

Sludge volume index (SVI): a process control calculation used to evaluate the settling quality of the activated sludge. Requires the SSV_{30} and mixed liquor suspended solids test results to calculate.

Soil moisture (soil water): water diffused in the soil. It is found in the upper part of the zone of aeration from which water is discharged by transpiration from plants or by soil evaporation.

Specific heat: the heat capacity of a material per unit mass. The amount of heat (in calories) required to raise the temperature of one gram of a substance 1°C; the specific heat of water is 1 calorie.

Storm sewer: a collection system designed to carry only stormwater runoff.

Stormwater: runoff resulting from rainfall and snowmelt.

Stream: a general term for a body of flowing water. In hydrology, the term is generally applied to the water flowing in a natural channel as distinct from a canal. More generally, it is applied to the water flowing in any channel, natural or artificial. Some types of streams are: (1) Ephemeral: a stream that flows only in direct response to precipitation, and whose channel is at all times above the water table. (2) Intermittent or Seasonal: a stream that flows only at certain times of the year when it receives water from springs, rainfall, or from surface sources such as melting snow. (3) Perennial: a stream that flows continuously. (4) Gaining: a stream or reach of a stream that receives water from the zone of saturation. An effluent stream. (5) Insulated: a stream or reach of a stream that neither contributes water to the zone of saturation nor receives water from it. It is separated from the zones of saturation by an impermeable bed. (6) Losing: a stream or reach of a stream that contributes water to the zone of saturation. An influent stream. (7) Perched: a perched stream is either a losing stream or an insulated stream that is separated from the underlying groundwater by a zone of aeration.

Supernatant: the liquid standing above a sediment or precipitate.

Surface tension: the free energy produced in a liquid surface by the unbalanced inward pull exerted by molecules underlying the layer of surface molecules.

Surface water: lakes, bays, ponds, impounding reservoirs, springs, rivers, streams, creeks, estuaries, wetlands, marshes, inlets, canals, gulfs inside the territorial limits of the state, and all other bodies of surface water, natural or artificial, inland or coastal, fresh or salt, navigable or nonnavigable, and including the beds and banks of all watercourses and bodies of surface water, that are wholly or partially inside or bordering the state or subject to the jurisdiction of the state. Waters in treatment systems that are authorized by state or federal law, regulation, or permit. and which are created for the purpose of water treatment, are not considered to be waters in the state.

T

Thermal pollution: the degradation of water quality by the introduction of a heated effluent. Primarily the result of the discharge of cooling waters from industrial processes (particularly from electrical power generation); waste heat eventually results from virtually every energy conversion.

Titrant: a solution of known strength of concentration; used in titration.

Titration: a process whereby a solution of known strength (titrant) is added to a certain volume of treated sample containing an indicator. A color change shows when the reaction is complete.

Titrator: an instrument (usually a calibrated cylinder (tube-form)) used in titration to measure the amount of titrant being added to the sample.

Total dissolved solids: the amount of material (inorganic salts and small amounts of organic material) dissolved in water and commonly expressed as a concentration in terms of milligrams per liter.

Total suspended solids (TSS): total suspended solids in water, commonly expressed as a concentration in terms of milligrams per liter.

Toxicity: the occurrence of lethal or sublethal adverse affects on representative sensitive organisms due to exposure to toxic materials. Adverse effects caused by conditions of temperature, dissolved oxygen, or nontoxic dissolved substances are excluded from the definition of toxicity.

Transpiration: the process by which water vapor escapes from the living plant—principally the leaves—and enters the atmosphere.

V

Vaporization: the change of a substance from a liquid or solid state to a gaseous state.

VOC (Volatile Organic Compound): any organic compound that participates in atmospheric photochemical reactions except for those designated by the USEPA Administrator as having negligible photochemical reactivity.

W

Waste activated sludge solids (WASS): the concentration of suspended solids in the sludge being removed from the activated sludge process.

Wastewater: the water supply of a community after it has been soiled by use.

Waterborne disease: a disease caused by a microorganism that is carried from one person or animal to another by water.

Water cycle: the process by which water travels in a sequence from the air (condensation) to the earth (precipitation) and returns to the atmosphere (evaporation). It is also referred to as the hydrologic cycle.

Water quality: a term used to describe the chemical, physical, and biological characteristics of water with respect to its suitability for a particular use.

Water quality standard: a plan for water quality management containing four major elements: water use, criteria to protect uses, implementation plans and enforcement plans. An anti-degradation statement is sometimes prepared to protect existing high quality waters.

Watershed: the area of land that contributes surface runoff to a given point in a drainage system.

Water supply: any quantity of available water.

Weir: a device used to measure wastewater flow.

Z

Zone of aeration: a region in the earth above the water table. Water in the zone of aeration is under atmospheric pressure and would not flow into a well.

Zoogleal slime: the biological slime that forms on fixed film treatment devices. It contains a wide variety of organisms essential to the treatment process.

Index

AA methods, *see* Atomic absorption methods
Achilles' heel, 114
Activated sludge process, 56
Active power, 78, 106
Active security barriers, 194–197
Adaptation, of benchmarking process, 15
Adapted BOD analyzer, 217
Address Verification Service, 175
Aeration blowers, 108
　anoxic and anaerobic zone mixing, 120–121
　high-speed gearless (turbo) blowers, 111–112
　hyperbolic mixing, 121–123
　innovative and emerging energy conservation measures, 113–114
　membrane bioreactors (MBRs), 119–120
　new diffuser technology, 112–113
　single-stage centrifugal blowers, 112
　UV disinfection, 114–119
Aerostrip diffuser, 113
Airport Water Reclamation Facility (AWRF), Prescott, Arizona (February 21, 2012), 84–85
Alarm system, 191–193, 206, 219, 224
Alternating current (A-C) motor, 101
Amendable Cyanide, 221
American Public Works Association (APWA), 8
American Society for Testing Materials, 60
American Society of Civil Engineers (ASCE), 60, 62
American Water Works Association (AWWA), 8, 61
Analysis, of benchmarking process, 15
Annual energy savings, 104
Annunciator, 192
Anoxic and anaerobic zone mixing, 120–121
Anti-intrusion detection system evasion techniques, 230
Antivirus and pest eradication software, 227–228
Antivirus offers, fake, 157
Apparent power, 106
APWA, *see* American Public Works Association
Arming station, 191, 192
Arsenic measurement system, 217
ASCE, *see* American Society of Civil Engineers
Association of Metropolitan Water Agencies, 8
Atomic absorption (AA) methods, 217
Attribute, defined, 3
AWRF, *see* Airport Water Reclamation Facility
AWWA, *see* American Water Works Association

Backflow prevention devices, 193–194
Backpressure, 194

Backsiphonage, 194
Barbed wire, 203
Baseline audit, 74–75
Baseline data, collection of, 70–73
BAT, *see* Best available technology
Benchmarking, 13–14, 69
　baseline audit, 74–75
　case study, 15–16
　collection of baseline data and tracking energy use, 70–73
　defined, 3
　equipment inventory and distribution of demand and energy, 75–76
　field investigation, 75
　potential results, 14
　process, 14
　steps, 15
BEP, *see* Best efficiency point
Best available technology (BAT), 28
Best efficiency point (BEP), 98
Bio^{M}x system, 122, 123
Biological nitrogen removal, 120
Biological nutrient removal (BNR), 120
Biometric security systems, 198
　biometric hand and finger geometry recognition, 199
　iris recognition, 199–200
Block power, 78
Blowers, 108–109
Blue Plains WWTP, 121–122
BNR, *see* Biological nutrient removal
Bollards, 196–197
Bowery Bay Water Pollution Control Plant, 121

Cantilever crash beams, 196
Cantilever gates, 196
Card identification/access/tracking systems, 200–202
Card reader system, 200–202
Cash Flow Opportunity (CFO) calculator, 105
CCCSD, *see* Central Contra Costa Sanitary District
CCTV system, *see* Closed circuit television system
Central Contra Costa Sanitary District (CCCSD), 119
Centrifugal blowers
　multi-stage, 109
　single-stage, 109, 112
CFO calculator, *see* Cash Flow Opportunity calculator

249

Chandler Municipal Utilities, Arizona (February 21, 2012), 83–84
Chemical sensors, 216
   adapted BOD analyzer, 217
   arsenic measurement system, 217
   chlorine measurement system, 219–220
   portable cyanide analyzer, 220–221
   total organic carbon (TOC) analyzer, 218–219
Chesapeake Bay and nutrients, 30–32
Chino Water Production Facility, Prescott, Arizona (February 21, 2012), 93–94
Chlorine measurement system, 219–220
Classic buffer overflow, 132, 181
Clean drinking water, 20, 92
Clean Water Act (CWA), 33, 62, 215
Clean Water Act Amendments, 28
Clean Water Needs Survey Report to Congress (1996), 56
Closed circuit television (CCTV) system, 191, 214–215
Cloud-based services, 158
College graduate, 4
Community sustainability, 9–10
Community water systems (CWSs), 57, 59, 185
Complex power, 106
Computer Crime and Security Survey, 140
Computer Security Institute and Federal Bureau of Investigation, 140
Configuration management processes, 149
Construction Grants program, 62
Contact tanks, 115
Control panel, 192
Control system, 230
   cyber threats to, 143, 232–233
   potential vulnerabilities in, 231
   securing, 144
Crash barriers, *see* Active security barriers
Crash beam barriers, 195, 196
Credibility, establishing, 6
Criminal groups, 129, 178, 179
Cross-site request forgery, 132, 181
Cross-site scripting, 132, 181
Cryptographic weakness, 132, 181
Customer satisfaction, 6, 9
CVV2 code, 175
CWA, *see* Clean Water Act
CWSs, *see* Community water systems
Cyanide, 220–221
Cyber criminals, 155, 156, 160–161
Cyber protection devices, 227
   antivirus and pest eradication software, 227–228
   firewalls, 228–229
   network intrusion hardware and software, 229–230
Cyber security program, 148–150
Cyber threats to control systems, 143, 232–233

Decisions making, 6
Defense-in-depth, 148–149, 164
Demand side management (DSM), *see* Energy demand management
Demilitarized zones (DMZs), 145
Denial-of-service attacks, 148, 160
Denitrification, 120
Department of Defense (DoD), 131, 178, 180
Detection devices, 191, 193
N, N-Diethyl-p-phenylenediamine (DPD) colorimetric method, 219, 220
Digital network security, 127–133, 176–178
Digital relays, 225
Disaster recovery plans, 149
DMZs, *see* Demilitarized zones
DoD, *see* Department of Defense
Door frames, 211, 212
Doorway security systems, 210
Double-cylinder locks, 207
DPD colorimetric method, *see* N, N-diethyl-p-phenylenediamine colorimetric method
Drinking water, clean, 20, 92
Drinking water capital stock, 63–65
Drinking water industry, characteristics of, 57–59
Drinking water reservoirs, 209
Drinking water treatment, 55–56
Drop-arm crash beams, 196
Dysfunctional management (case study), 38–40
   defining, 41–46

Eastern Municipal Water District (EMWD), California (February 21, 2012), 87–88
ECMs, *see* Energy conservation measures
Education, 5
Effective management, regulatory view of, 8
   attributes of effectively managed water sector utilities, 8
   community sustainability, 9–10
   customer satisfaction, 9
   employee and leadership development, 10
   financial viability, 9
   infrastructure stability, 9
   operational optimization, 9
   operational resiliency, 9
   product quality, 9
   stakeholder understanding and support, 10
   water resource adequacy, 10
Effective public works manager, 4, 5
Effective utility management, defined, 3
Electrical load management, 77
   energy demand management, 78
   energy management strategies, 78–79
   rate schedules, 78
   success stories, 79–82
Electric Power Research Institutes (EPRI), 75
Electromechanical relays, 225–226

Index

Electronic controllers, 224–226
Email, 165
    developing email usage policy, 166
    protection of sensitive information sent via, 166
    sensible email retention policy, setting, 166
    setting up a spam email filter, 165
    training employees in responsible usage of, 165
Employee and leadership development, 10
Employee internet usage policy, establishing, 154
Employees, 169
    developing a hiring process that properly vets candidates, 169
    performing background checks and credentialing, 169–170
    providing security training for, 171–172
    setting appropriate access controls for, 170
    taking care in dealing with third parties, 170
EMWD, *see* Eastern Municipal Water District
End of lamp life, 118
Energy audit, 74–75
Energy conservation measures (ECMs), 97
    aeration blowers (*see* Aeration blowers)
    power factor, 106
    pumping system design, 99–101
    pump motors, 101–105
    variable frequency drives (VFDs), 106–180
Energy demand management, 78
Energy efficient operating strategies, 77
    electrical load management (*see* Electrical load management)
    operation and maintenance, 82
Energy management program, 72
Energy management strategies
    energy survey, conducting, 78–79
    load to off-peak, shifting, 79
    peak demand, reducing, 79
    power factor, improving, 79
Energy Policy Act (EPACT) of 1992, 102–103
Energy provider, 72
ENERGY STAR® Cash Flow Opportunity (CFO) calculator, 105
Energy use, tracking, 70–73
Enterprise Credit Union, 128, 177
Environmental Protection Agency (EPA), 186
    Construction Grants program, 62
    Drinking Water Infrastructure Needs Survey (2007), 63
EPA, *see* Environmental Protection Agency
EPACT, *see* Energy Policy Act of 1992
EPRI, *see* Electric Power Research Institutes
Equipment inventory and distribution of demand and energy, 75–76
Extensible Markup Language (XML), 141, 164, 230

Fayol's Five Functions of Management, 4
FCC, *see* Federation Communication Commission
Federal Energy Regulatory Commission (FERC), 231
Federation Communication Commission (FCC), 226
Fences, 202–203
FERC, *see* Federal Energy Regulatory Commission
Field investigation, 75
Films for glass shatter protection, 203–204
Finance, ratepayers' rates, and regulations, 6
Financial viability, 9
Finger geometry recognition, 199
Fire alarm systems, 192
Fire detection/fire alarm systems, 193
Fire hydrant locks, 204–205
Firewalls, 228–229
Foreign intelligence services, 129, 178
Frame grabber, 200
Free cyanide, 220–221
Future pricing options, 78
F. Wayne Hill Water Resources Center in Gwinnett County, Georgia, 122

Galileo's theory, 20
Gap analysis, defined, 3
Gates, 196
General Electric (GE), 119
Gestapo mentality, 5
GE ZENON ZeeWeed MBR, 120
Glass shatter protection, films for, 203–204
Great Sanitary Awakening, 56
Groundwater sources, 57
Groundwaters under direct influence (GWUDI), 21

Hackers, 129, 178, 179
Hacktivism, 129, 178
Hacktivists, 129, 178, 179
Hand and finger geometry recognition, 199
Hardwired systems, 193
Hatch security, 205
Hawaii County Department of Water Supply (February 21, 2012), 86–87
HCN, *see* Hydrogen cyanide
Heap-based buffer overflow, 133, 182
Hidden function, 26, 33
High-speed gearless blowers, 111–112
HMI, *see* Human Machine Interface
Homeland Security, 189
HTTP content, *see* Hypertext transfer protocol content
Huber system, 119
Human Machine Interface (HMI), 73, 135
Hydrant lock, 204–205
Hydrofoil mixer, 121–122
Hydrogen cyanide (HCN), 220, 221
Hyperbolic mixing, 121–123

252 Index

HYPERCLASSIC HC RKO 2500, 121
Hyperclassic mixer, 121
Hypertext transfer protocol (HTTP) content, 164

ICS-CERT, *see* Industrial Control Systems Cyber Emergency Response Team
IDSs, *see* Intrusion detection systems
Independent system operator (ISO), 145
Industrial Control Systems Cyber Emergency Response Team (ICS-CERT), 128, 177
Information Assurance Task Force of the National Security Telecommunications Advisory Committee, 232
Information warfare, 129, 178
Infrastructure, 50
    maintaining, 28–29
    stability, 9
Inside threat, 129, 178
Insufficient authentication requirements, 132, 181
Integer overflow, 133, 182
Intel 8088 processor, 142, 231
Interior intrusion sensors, 193, 206
Internal network, 158
Internal trend analysis, defined, 3
Internet usage policy of employee, establishing, 154
Interruptible rates, 78
Intrusion detection systems (IDSs), 145
Intrusion sensors, 192, 206
    buried exterior intrusion sensors, 206
Iris recognition systems, 199–200
ISO, *see* Independent system operator
IT security action items, 153
    email, 165–166
    employees (*see* Employees)
    mobile devices, 167–169
    network security, 158–160
    operational security (OPSEC), 172–174
    payment cards, 174–175
    policy development and management, 154–155
    scams and fraud, 155–157
    web server security (*see* Web server security)

Key-logging malware, 157

Ladder access control, 206–207
LAN, *see* Local area network
Leadership, 7, 10
Learning process, 14
Life-cycle cost, defined, 3
Life of an asset, 59–60
Linear gate, 196
Local alarm, 192
Local area network (LAN), 227
Local only alarm, 192

Locks, 207
    fire hydrant locks, 204–205

Malicious insiders, 179
Malicious outsiders, unknown, 179
Malicious software, developing a layered approach to guard against, 157
Malware, 227
    protection against, 157
Managing Total Quality (MTQ), 45
Manhole intrusion sensors, 207–208
Manhole locks, 208
Mars, 37
Master Terminal Unit (MTU), 73, 135
Maximum contaminant levels (MCLs), 186, 221
MBRs, *see* Membrane bioreactors
MCC, *see* Motor control center
MCLs, *see* Maximum contaminant levels
Membrane bioreactors (MBRs), 119–120
Membrane fouling, 119
Metric Million British Thermal Units (MMBTU), 70
MLE process, *see* Modified Ludzack-Ettinger process
MMBTU, *see* Metric Million British Thermal Units
Mobile devices, 167
    employing strategies for email, texting, and social networking, 168–169
    encrypting the data on, 168
    ensuring all devices are wiped clean prior to disposal, 169
    having users password protect access to, 168
    making sure all software is up to date, 168
    setting reporting procedures for lost or stolen equipment, 169
    top threats targeting, 167–168
    urging users to be aware of their surroundings, 168
    using security software on all smartphones, 168
Modified Ludzack-Ettinger (MLE) process, 120
Monthly energy bills, 72
Motor control center (MCC), 75
Motor Decisions Matter campaign, 105
Motor efficiency and efficiency standards, 102–103
Motor management programs, 103–105
MotorMaster+, 105
MTQ, *see* Managing Total Quality
MTU, *see* Master Terminal Unit
Multiple-barrier approach, 20, 21, 25
Multiple-barrier concept, 24–27
Multi-stage centrifugal blowers, 109

NACWA, *see* National Association of Clean Water Agencies
National Association of Clean Water Agencies (NACWA), 8, 67

Index

National Association of Water Companies (NAWC), 8
National Electrical Manufacturers Association (NEMA), 103
National Environmental Policy Act (NEPA), 28
National Fire Protection Association (NFPA) 72, 192
National Infrastructure Protection Center (NIPC), 143, 232
National Pollutant Discharge Elimination System (NPDES), 28, 216
National Primary Drinking Water Regulation, 27
National Secondary Drinking Water Regulation, 27
Nations, 179
NAWC, see National Association of Water Companies
NCWSs, see Noncommunity water systems
NEMA, see National Electrical Manufacturers Association
NEPA, see National Environmental Policy Act
Network honeypot, 229
Network infrastructure, employing, 164
Network intrusion hardware and software, 229–230
Network security
  cloud-based services, 158
  internal network, 158
  organization Wi-Fi, securing and encrypting, 159
  regularly updating all applications, 159–160
  remote access, 160
  safe web browsing rules, setting, 160
  sensitive company data, encrypting, 159
  strong password policies, developing, 159
New diffuser technology, 112–113
New management graduate, 4–5
New prospective manager, 5
NFPA 72, see National Fire Protection Association 72
NIPC, see National Infrastructure Protection Center
Nitrification, 120
Noncommunity water systems (NCWSs), 185
NPDES, see National Pollutant Discharge Elimination System

O&M capital stock, see Operating and maintaining capital stock
Observation, of benchmarking process, 15
Occupational Safety and Health Administration (OSHA), 29, 114
Online fraud, protection against, 156
On-the-job training, 5
OpenPGP standard, 159
Open redirect, 133, 182
Operating and maintaining (O&M) capital stock, 60–61

Operating revenue, defined, 3
Operating system command injection, 132, 181
Operational optimization, 9
Operational resiliency, 9
Operational security (OPSEC), 172
  critical information, identity of, 172–173
  risk, assessing, 174
  threats, analyzing, 173
  vulnerabilities, analyzing, 173
Operations and maintenance expenditure, defined, 3
OPSEC, see Operational security
Organization effectiveness, measuring, 10–11
  benchmarking, 13–15
  bottom line on measurement, 17
Organization's security requirements, 162
ORP measurements, see Oxidation-reduction potential measurements
OSHA, see Occupational Safety and Health Administration
OTE, see Oxygen transfer efficiency
Oxidation-reduction potential (ORP) measurements, 122
Oxygen transfer efficiency (OTE), 113

Pacific Gas and Electric (PG & E) company, 116
Panel diffusers, 113
PAOs, see Phosphate accumulating organisms
Paradigm, 20–22
Paradigm shift, defined, 21
Passive security barriers, 197–198
Password policies, developing, 159
Password protection on mobile devices, 168
Passwords, complex, 231
Path traversal, 132, 182
Payment cards, 174
  controlling access to payment systems, 175
  evaluating whether the company needs to keep all the data it has stored, 174–175
  understanding and cataloging customer and card data you keep, 174
  using secure tools and services, 175
  using security tools and resources, 175
PDA device, see Personal digital assistant device
PDWTP, see Port Drive Water Treatment Plant
Performance measurement, 3, 17
Perimeter intrusion sensors, 192
Personal digital assistant (PDA) device, 131
Pests, 228
PG & E company, see Pacific Gas and Electric company
Phishing, protection against, 156–157
Phishing and spear phishing, 132, 181
Phosphate accumulating organisms (PAOs), 120
Pipe networks, 62
Planning, of benchmarking process, 15
PLCs, see Programmable logic controllers

Pollution, prevention of, 28
Portable cyanide analyzer, 220–221
Portable/removable barriers, 197
Port Drive Water Treatment Plant (PDWTP), Lake Havasu, Arizona (February 21, 2012), 88–90
Positive displacement blowers, 109
Potable water, 19
Potential reputation risks, identifying, 155
Potential vulnerabilities in control systems, 231
POTWs, *see* Publicly owned treatment works
Power factor, 106
Power factor charges, 78
Pretexting, 156
Privatization
 and re-engineering, 32–33
 working against, 13
Process Safety Management (PSM) regulation, 29
Product quality, 9
Professional management, technical management versus, 37–41
Programmable logic controllers (PLCs), 73, 112, 122, 135, 225–226
Proprietary protocols, 146
Prospective manager, 5
Protocol analysis, 229
PSM regulation, *see* Process Safety Management regulation
Publicly owned treatment works (POTWs), 52, 55, 56, 97, 186
Public speaking skills, 7
Public utilities manager
 effective manager, 5
 general and daily tasks, 1, 2
 important functions within the organization, 5
 main areas of focus, 1, 2
 qualifications, 4–7
Public works management, 1
Pulsed large bubble mixing, 122–123
Pumping system design, 99–101
Pump motors, 101–102
 innovative and emerging technologies, 105
 motor efficiency and efficiency standards, 102–103
 motor management programs, 103–105
Pure water, 34

Qualifications of public utilities manager, 4–7

Rachel's Creek Sanitation District, 16–17
Radiation detection equipment, 222–223
Radiation detection equipment for monitoring personnel and packages, 208
Radios, two-way, 226
Ratepayers' rates, 6
Reactive power, 78, 106

Real-time pricing, 78
Red Teams, 147
Re-engineering, 32–33
Remote access, 160
Remote Terminal Unit (RTU), 73, 135
Reputation risks, types of, 155
Research, of benchmarking process, 15
Reservoir covers, 209–210
Resource planning, 6–7
Retired On Active Duty, 33
Retractable bollards, 197
Risk Management Planning (RMP) regulation, 29
Rotor, 101–102
RTU, *see* Remote Terminal Unit

Safe Drinking Water Act (SDWA), 24, 33, 215, 217, 219
Safe Drinking Water Act Arsenic Rule, 217
Safe Drinking Water Act Reauthorization, 24
Safe web browsing rules, setting, 160
SCADA systems, *see* Supervisory Control and Data Acquisition systems
Scams and fraud, 155
 fake antivirus offers, 157
 malicious software, developing a layered approach to guard against, 157
 malware, protection against, 157
 online fraud, protection against, 156
 phishing, protection against, 156–157
 social engineering, training employees to recognize, 156
 telephone information seekers, verifying the identity of, 158
Scareware, 157
SDWA, *see* Safe Drinking Water Act
Secure Socket Layer (SSL) certificate, 159
Securing control systems, 144
Security barriers, 198
Security doors, 212
Security hardware devices, 190
 communication and integration, 224–227
 cyber protection devices, 227–230
 cyber threats to control systems, 232–233
 known vulnerabilities, adoption of technologies with, 230–232
 physical asset monitoring and control devices
  aboveground, outdoor equipment enclosures, 190–191
  active security barriers, 194–197
  alarm system, 191–193
  backflow prevention devices, 193–194
  biometric security systems, 198–200
  card identification/access/tracking systems, 200–202
  fences, 202–203
  films for glass shatter protection, 203–204

Index

fire hydrant locks, 204–205
hatch security, 205
intrusion sensors, 206
ladder access control, 206–207
locks, 207
manhole intrusion sensors, 207–208
manhole locks, 208
passive security barriers, 197–198
radiation detection equipment for monitoring personnel and packages, 208
reservoir covers, 209–210
side-hinged door security, 210–212
valve lockout devices, 212–213
vent security, 213–214
visual surveillance monitoring, 214–215
securing control systems, 233
water monitoring devices (*see* Water monitoring devices)
Security requirements, of organization, 162
Security roles and responsibilities, establishing, 154
Self-assessment processes, 149
Senior organizational leadership, 150
Sensitive company data, encrypting, 159
Sensors, *see* Detection devices
September 11, 2001, consequences of, 186–190
Side-hinged door security, 210–212
Siemens, 119
SIM card, *see* Subscriber Identification Module card
Single-cylinder locks, 207
Single-stage centrifugal blowers, 109, 112
Smartphones, using security software on, 168
Social engineering, 150, 156
Social media policy, establishing, 154–155
Sodium iodide (NaI) scintillator, 223
Solids residence time (SRT), 120
Soluble Cyanide, 221
Somerton Municipal Wastewater Treatment Plant, Arizona (February 21, 2012), 94–95
Somerton Municipal Water, Arizona (February 21, 2012), 85–86
Spam email filter, setting up, 165
Spyware, 227
SQL injection, *see* Structured Query Language injection
SRT, *see* Solids residence time
SSL certificate, *see* Secure Socket Layer certificate
Staffing wastewater treatment plants, 7
Stakeholder relations, 6
Stakeholder understanding and support, 10
Standard operating procedure, defined, 3
Stator, 101
Stewardship, defined, 3

Stickiness, 34
Strategic planning, 3, 5
Structured Query Language (SQL) injection, 132, 181
Subscriber Identification Module (SIM) card, 168
Success, accomplishment of, 5
Supervisory Control and Data Acquisition (SCADA) systems, 72–74, 135–136, 226
applications, 136–138
in water/wastewater system, 73–74
steps to improve SCADA security, 144
clearly define cyber security roles, responsibilities, and authorities, 147–148
clearly identify cyber security requirements, 149
conduct physical security surveys and assess all remote sites, 147
conduct routine self-assessments, 149
disconnect unnecessary connections to SCADA network, 145
document network architecture and identify systems that serve critical functions, 148
establish a network protection strategy based on the principle of defense-in-depth, 148–149
establish a rigorous, ongoing risk management process, 148
establish effective configuration management processes, 149
establish policies and conduct training, 150
establish Red Teams, 147
establish strong controls over any medium that is used as a backdoor into SCADA network, 146
establish system backups and disaster recovery plans, 149
evaluate and strengthen the security of any remaining connections to SCADA networks, 145
harden SCADA networks by removing or disabling unnecessary services, 145–146
identify all connections to SCADA networks, 144–145
implement internal and external intrusion detection systems and establish 24-hours-a-day incident monitoring, 146–147
implement the security features provided by device and system vendors, 146
perform technical audits of SCADA devices and networks to identify security concerns, 147
proprietary protocols, 146

Index

senior organizational leadership, 150
vulnerabilities
   adoption of technology with known vulnerabilities, 141–143
   cyber threats to control systems, 143
   risk, increasing, 141
   securing control systems, 144
   U.S. energy grid, spies hacked into, 138–141
Surface water sources, 57
Surface Water Treatment Rule (SWTR), 21
Sustainability, defined, 4, 49
Sustainable development, defined, 51
Sustainable energy future, planning for, 67
   benchmarking, 69–76
   wastewater and drinking water treatment, 67–69
Sustainable water/wastewater infrastructure, 51–53
Swarming attacks, 143
Swing gates, 196
SWTR, *see* Surface Water Treatment Rule
System backups, 149

Technical management versus professional management, 37–41
Telephone information seekers, verifying the identity of, 158
TEP, *see* Tucson Electric Power
Terrorists, 179
Threats, analyzing, 173
Time-of-use rates, 78
TMWA, *see* Truckee Meadows Water Authority
TOC analyzer, *see* Total organic carbon analyzer
Total cyanide, 220, 221
Total organic carbon (TOC) analyzer, 218–219
Total Quality Management (TQM), 13, 45
Toxicity monitoring/toxicity meters, 224
TQM, *see* Total Quality Management
Traffic anomaly detection, 229
Triple Bottom Line scenario, 50
Truckee Meadows Water Authority (TMWA), Reno, Nevada (February 21, 2012), 90–92
True power, *see* Active power
Trusted third parties, 132, 181
Tucson Electric Power (TEP), 92
Tucson Water, Arizona (February 21, 2012), 92–93
Turbo blowers, *see* High-speed gearless blowers
Two-way radios, 226

Ultra-Efficient and Power-Dense Electric Motors, 105
Uncontrolled format string, 133, 182
United States Environmental Protection Agency (USEPA), 8, 21, 24, 25, 27–29
Unrestricted upload of files, 133, 182
Un-trusted sphere, inclusion of functionality from, 133, 182
U.S. Army Corps of Engineers, 60
U.S. Department of Energy (USDOE), 105
USEPA, *see* United States Environmental Protection Agency
UV disinfection, 114
   design, 116
      lamp selection, 116–117
      pretreatment, 116
      system hydraulics to promote mixing, 117
      system turndown, 117
   operation and maintenance, 117
      automation, 117–118
      lamp cleaning and replacement, 118–119
UV transmittance (UVT), 115

Valve lockout devices, 212–213
Variable frequency drives (VFDs), 98, 106
   applications, 107
   energy savings, 107
   strategies, for wastewater pumping stations, 107–108
Vent security, 213–214
VFDs, *see* Variable frequency drives
Video camera, 200
Virtual Private Network (VPN) system, 160
Viruses, 228
Virus writers, 129, 178
Visual surveillance monitoring, 214–215
Volatile organic compounds (VOCs), 220, 222
VPN system, *see* Virtual Private Network system
Vulnerabilities, analyzing, 173

WAD Cyanide, *see* Weak Acid Dissociable Cyanide
Waldrop Syndrome (case study), 11–13
WAP, *see* Wireless Access Point
Wastewater and drinking water industries, characteristics of, 55–59
   capital stock and impact on operations/maintenance, 59
      operating and maintaining (O&M) capital stock, 60–61
      useful life of assets, 59–60
      wastewater capital stock, 61–63
      drinking water capital stock, 63–65
      drinking water treatment, 55–56
      wastewater treatment, 55
      wastewater treatment process, 56–57
Wastewater and drinking water treatment, 67–69
Wastewater capital stock, 61–63
Wastewater industry, characteristics of, 56
Wastewater pumping stations, VFD strategies for, 107–108
Wastewater service professionals, 26

Index

Wastewater treatment, 55, 186
    management, 37
    process, 56–57
Wastewater treatment plants (WWTPs), 52, 98, 107, 112–114, 116–120
Wastewater utilities, 103
Water, 26, 33, 34
    radiation detection equipment for monitoring water assets, 222–223
    technical management versus professional management, 37–41
    use, 35–37
    and wastewater treatment services, 26
Water and wastewater treatment operations, management issues in, 19
    compliance with new, changing, and existing regulations, 27–28
    infrastructure, maintaining, 28–29
    multiple-barrier concept, 24–27
    paradigm shift, 20–22
    privatization and/or re-engineering, 32–33
Water Environment Federation (WEF), 8, 60, 102
Water Infrastructure Gap Analysis, 53
Watering hole, 132, 181
Water monitoring devices, 215
    chemical sensors (*see* Chemical sensors)
    radiation detection equipment, 222–223
    sensors for monitoring chemical, biological, and radiological contamination, 216
    toxicity monitoring/toxicity meters, 224
    volatile organic compounds (VOCs), portable field monitors to measure, 222
Water Pollution Control Act Amendments, 28
Water Pollution Control Plant (WPCP), 121
Water Protection Task Force, 186
Water resource adequacy, 10
Watershed health, defined, 4
Water treatment operations management, 37
Water treatment plants (WTPs), 52
Water/wastewater (W/WW) plants, 52
Water/wastewater infrastructure, 49
    sustainable, 51–53
Water/wastewater infrastructure gap, 53
    energy efficiency, 53–54
Water's stickiness, 34
Weak Acid Dissociable (WAD) Cyanide, 220, 221

Web server security, 160
    carefully planning and addressing, 161
    committing to an ongoing process of maintaining, 165
    content published on the organization's website, 162–163
    implementing appropriate security management practices and controls, 161–162
    network infrastructure, employing, 164
    using active content judiciously after balancing the benefits and risks, 164
    using authentication and cryptographic technologies, 164
    web content protection from unauthorized access or modification, 163
    web server application meeting the organization's security requirements, 162
    web server systems meeting the organization's security requirements, 162
Wedge barriers, 195
WEF, *see* Water Environment Federation
WET tests, *see* Whole effluent toxicity tests
Whole effluent toxicity tests (WET tests), 224
Wi-Fi, securing and encrypting, 159
Window and Door Manufacturers Association, 210–211
Wireless access control, 159
Wireless Access Point (WAP), 226
Wireless data communications, 226–227
Wireless encryption, 159
Wireless local area network (WLAN), 159
Wireless Network Interface Card, 226, 227
Wireless systems, 193
Wire-to-water efficiency, 98
WLAN, *see* Wireless local area network
WPCP, *see* Water Pollution Control Plant
WTPs, *see* Water treatment plants
WWTPs, *see* Wastewater treatment plants
W/WW plants, *see* Water/wastewater plants

XML, *see* Extensible Markup Language

Yuck Factor, 7

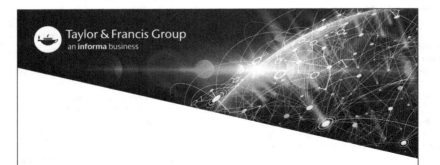